The Pilot's Travel & Recreation Guide

TRAVEL & RECREATION GUIDE

Northeast United States and Eastern Canada

DOUGLAS S. CARMODY

The Pilot's Travel & Recreation Guide

Northeast and Eastern Canada

Douglas S. Carmody

McGraw-Hill

New York   San Francisco   Washington, D.C.
Auckland   Bogotá   Caracas   Lisbon   London   Madrid   Mexico City
Milan   Montreal   New Delhi   San Juan   Singapore   Sydney   Tokyo   Toronto

McGraw-Hill

A Division of The McGraw·Hill Companies

1 2 3 4 5 6 7 8 9 0   DOC/DOC   9 0 4 3 2 1 0 9

ISBN 0-07-001743-3

The sponsoring editor for this book was Shelley Ingram Carr, the editing supervisor was Penny Linskey, and the production supervisor was Clare Stanley. Interior design by Jaclyn J. Boone. It was set in Weidemann Book by Michele Pridmore of McGraw-Hill's Professional Book Group composition unit in Hightstown, N.J.

Printed and bound by R. R. Donnelley & Sons Company.

This book is printed on recycled, acid-free paper containing a minimum of 50% recycled, de-inked fiber.

For Irene Carmody
(Thanks again to Shirley Kumpf and Pam Bratz for their encouragement and typing skills in completing this project.)

Contents

Introduction   ix
Using This Guide   x

Introduction

Traveling by light airplane reveals vistas and avenues of excitement unknown to earthbound wanderers. Weekend getaways take on a whole new meaning as we travel distances that would use up the whole week if we had to drive. This guide is an expression of that freedom. The hotels, restaurants, and attractions have all been chosen by pilots for pilots, but it's a continuing quest. Please feel free to write to me to update this guide or share your experiences at listed locations. No one has paid any fee to be listed. I'm sure I've overlooked some great places and restaurants, but it is by shear oversight, not intent.

Douglas S. Carmody

Using This Guide

First, find the state and the city in the Contents. Each listing includes a condensed version of the airport facility directory complete with a diagram of the airport. If a city has more than one airport, they are listed alphabetically.

Transportation

I have listed rental cars, taxis, and courtesy cars. Using a courtesy car is unreliable. They break down or are unavailable, so please check before you go. Remember to put gas in the FBO's car after you use it. After all, you used the car for free.

Area Attractions

I have listed things I thought pilots and their families would be interested in. Let me know what you think.

Interesting Facts and Events

Here's a calendar of interesting events listed for the city.

Lodging

I have tried to give you a cross-section in quality and price. All listed hotels and motels are within 30 minutes of the primary airport. **Bed and Breakfasts** are listed when appropriate.

Restaurants

From hamburgers at the airport grill to five-star restaurants.

Feel free to send your favorite place for the next edition to:

Doug Carmody
39A Airport Circle
Beaufort, SC 29902

Have a great flight!

The
Pilot's Travel
&
Recreation Guide

Connecticut

Bridgeport

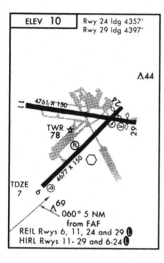

ELEV 10	Rwy 24 ldg 4357' Rwy 29 ldg 4397'

Λ44

4761 X 150

TWR
78

4677 X 150

TDZE
7

69
060° 5 NM
from FAF
REIL Rwys 6, 11, 24 and 29 ◖
HIRL Rwys 11- 29 and 6-24 ◖

Bridgeport is a manufacturing city with many well-known companies producing a wide array of products. It is also the home of the University of Bridgeport, founded in 1927. P. T. Barnum (Greatest Show on Earth) was a mayor of Bridgeport, and residents grew accustomed to the circus atmosphere that accompanied the Barnum administration.

Airport Igor I. Sikorsky Memorial (BDR) is located 3 miles southeast of the city. Coordinates: N41° 09.81' W073° 07.57'.

Traffic Pattern The traffic pattern is flown at 1010 ft MSL. Lefthand pattern for Runways 6, 11, 24, and 29. Runways 11 and 29 are asphalt surfaced, 4761 ft long by 150 ft wide. Runways 6 and 24 are asphalt surfaced, 4677 ft long by 150 ft wide.

FBO Bridgeport Air Center. (203) 375-3329. Hours: 6 A.M.–9 P.M. Frequency 123.0.

FBO Million Air–Bridgeport. (203) 377-1100. Hours: 6 A.M.–9 P.M. Frequency 122.95.

FBO Three Wing Corporation of Connecticut. (203) 375-5795. Hours: 7 A.M.–3:30 P.M. daily. Frequency 122.95.

Navigational Information BDR is located on the New York Sectional Chart or L25, L28 low en route chart. From the HVN 109.8 VOR 255° at 12.4 miles.

Instrument Approaches   ILS, VOR, and GPS approaches are available.

Fuel   100LL and JetA are available. Air BP, American Express, Amoco, and most major credit cards are accepted.

Frequencies
TOWER 120.9   GROUND 121.9   CLEARANCE DELIVERY 121.75
APPROACH NEW YORK 126.95   UNICOM 123.0   CTAF 120.9
ATIS 119.15.

Runway Lights   Pilot-controlled lighting. Tower beacon dusk to dawn.

Transportation

Rental Cars   Available at the terminal.
Avis, (203) 377-7390
Hertz, (203) 378-0513

Taxicabs
Suburban Limo, (203) 377-8294

Courtesy Cars   None Available.

Area Attractions

The Barnum Museum.   820 Main St. (203) 331-9881. An interesting museum that houses memorabilia from P. T. Barnum's life and circus career.

Beardsley Zoological Gardens.   Noble Ave., Beardsley Park, Off I-95. (203) 576-8082. The state's only zoo. Located on 30 acres, it houses more than 200 animals. Admission charged.

Discovery Museum.   4450 Park Ave. (203) 372-3521. In addition to a children's museum, the main building offers hands-on science and art exhibits, films, Challenger Learning Center, demonstrations, and a planetarium.

Other Attractions

Port Jefferson Ferry.   (203) 367-3043. Ride the ferry across Long Island Sound. Admission charged.

Interesting Facts & Events

Early July. *Barnum Festival.* Celebrates the life of showman P. T. Barnum. (203) 367-8495.

For More Information Contact the Bridgeport Chamber of Commerce at 10 Middle St., 06604; (203) 335-3800 for more information, a tourist package, and discounts.

Lodging

Unless otherwise noted, all lodging is within 30 minutes of the primary airport.

Holiday Inn
1070 Main St.
(203) 367-1985
Rates: $89–$129

Marriott Trumbull
180 Hawley Ln. Trumbull
(203) 378-1400
Rates: $164–$209

Restaurants

Black Rock Castle
2895 Fairfield Ave.
(203) 336-3990
Irish menu specializing in flaming Irish whiskey steak.
Dinner: $10–$18

Groton/New London

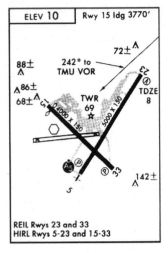

```
ELEV 10      Rwy 15 ldg 3770'

                      72±
88±      242° to
         TMU VOR
  86±                         TDZE
68±           TWR              8
              69

REIL Rwys 23 and 33
HIRL Rwys 5-23 and 15-33
```

Groton is home to the U.S. Navy's submarine force. The first diesel-powered submarine was built here in 1912, and the first nuclear-powered submarine, *Nautilus*, in 1955. New London is considered one of the best deep-water ports on the Atlantic coast.

Airport Groton-New London (GON) is located 3 miles southeast of the city. Coordinates: N41° 19.80' W072° 02.71'.

Traffic Pattern The traffic pattern is flown at 1010 ft MSL light aircraft, 1510 ft MSL turbine aircraft. Lefthand pattern for Runways 5, 15, 23, and 33. Runways 15 and 33 are asphalt surfaced, 4000 ft long by 150 ft wide. Runways 5 and 23 are asphalt surfaced, 5000 ft long by 150 ft wide.

FBO Columbia Air Services, Inc. (860) 449-1257. Hours: 5 A.M.–9 P.M., 24 hours on request. Frequency 122.95.

Navigational Information GON is located on the New York Sectional Chart or L25, L28 low en route chart.

Instrument Approaches ILS and VOR approaches are available.

Cautions There is intensive flight training in the area.

Fuel 100LL and JetA are available. Exxon and most major credit cards are accepted.

Frequencies
TOWER 125.6   GROUND 121.65   CLEARANCE DELIVERY 121.65
APPROACH CONTROL PROVIDENCE 125.75   BOSTON CENTER
124.85   UNICOM 122.95   CTAF 125.6   ATIS 127.0.

Runway Lights   Operated by the tower.

Transportation

Rental Cars   Available at the terminal.
Avis, (860) 445-8585
Budget, (860) 446-1915
National, (860) 445-7435

Taxicabs
Available at the terminal.

Courtesy Cars   The FBO has one available.

Area Attractions

Fort Griswold Battlefield State Park.   Monument St. and Park Ave.
(860) 449-6877. A tribute to Revolutionary solders who were killed in
1781 by British troops under the command of Benedict Arnold.

U.S.S. Nautilus Memorial/Submarine Force Library and Museum.
12 Crystal Lake Rd. (860) 449-3174 or (800) 343-0079. The world's first
nuclear-powered submarine is housed here.

New London
Joshua Hempsted House.   11 Hempsted St. (860) 443-7949. Oldest
house in the city (1678). Admission charged.

Lyman Allyn Art Museum.   625 Williams St. (860) 443-2545. View
displays of Colonial silver, 18th- and 19th-century furniture, dolls, and
dollhouses. Admission charged.

Other Attractions

Oceanographic Cruise.   Take an educational cruise on Long Island
Sound. (800) 364-8472.

New London
Ocean Beach Park.   On Ocean Ave., Long Island Sound. (800) 510-
7263. Enjoy diversified water entertainment: ocean or pool swimming
and waterslide; concessions, boardwalk, and arcade. Admission charged.

Trumbull Golf Course.   119 High Rock. (860) 445-7991. 18 holes, par 73. Green fees.

U.S. Coast Guard Academy.   Mohegan Ave. (860) 444-8270. Review a cadet parade and explore the U.S. Coast Guard Museum. Visit the training ship *Eagle,* a 295-ft barque. Free.

Interesting Facts & Events

April.   *Connecticut Storytelling Festival.* Connecticut College. (860) 439-2764.

July.   *Sail Festival.* City Pier, New London. (860) 443-8331.

For More Information   Contact the Southeastern Connecticut Tourism District at 470 Bank St., New London, (860) 444-2206 or (800) 863-6569, for more information, a tourist package, and discounts.

Lodging

Unless otherwise noted, all lodging is within 30 minutes of the primary airport.

Groton

Best Western
360 Rt. 12
(800) 528-1234
Rates: $100–$125

Clarion Inn
156 Kings Hwy.
(860) 446-0660
Rates: $76–$124

Econolodge
425 Bridge St.
(860) 445-6550
Rates: $75–$125

Quality Inn
404 Bridge St.
(860) 445-8141
Rates: $90–$115

New London

Connecticut Yankee
I-95, Exit 74
(860) 739-5483
Rates: $38–$85

Lighthouse Inn
6 Guthrie Pl.
(860) 443-8411
Rates: $125–$200

Niantic Inn
345 Main St.
(860) 739-5451
Rates: $135

Red Roof Inn
707 Colman St.
I-95, Exit 82A
(860) 444-0001
Rates: $35–$77

Bed & Breakfast

Queen Anne Inn
265 Williams St.
(860) 447-2600
Rates: $84–$175

Sojourner Inn
Rt. 184
(202) 445-1986
Rates: $85–$200

Restaurants

Constantine's
252 Main St., Niantic
(860) 739-2848
Specializes in fresh seafood.

Ye Olde Taverne Steak and Chop
House
345 Bank St.
(860) 442-0353
Specializes in filet mignon.

New Haven

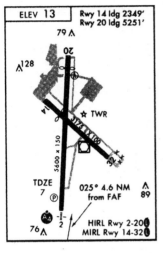

ELEV 13	Rwy 14 ldg 2349' Rwy 20 ldg 5251'

79 ∧

∧128

82 ∧

☆ TWR

3176 x 100

5600 x 150

TDZE 7 ℗

025° 4.6 NM from FAF ∧ 89

Ⓐ4

76 ∧

HIRL Rwy 2-20 ● MIRL Rwy 14-32 ●

Only 75 miles from New York City, New Haven is a world apart. This typical New England city boasts interesting Colonial history at almost every corner. Ivy League's Yale University makes New Haven one of the world's cultural centers.

Airport Tweed-New Haven (HVN) is located 3 miles southeast of the city. Coordinates: N41° 15.83' W072° 53.23'.

Traffic Pattern The traffic pattern is flown at 1014 ft MSL. Lefthand pattern for Runways 2, 14, 20, and 32. Runways 14 and 32 are asphalt surfaced, 3176 ft long by 100 ft wide. Runways 2 and 20 are asphalt surfaced, 5600 ft long by 150 ft wide.

FBO Robinson Aviation, Inc. (203) 457-9555. Hours: 7:30 A.M.–8:30 P.M., open for fueling at 6 A.M. Frequency 122.95.

Navigational Information HVN is located on the New York Sectional Chart or L25, L28 low en route chart. The HVN 109.8 VOR is on the field.

Instrument Approaches ILS, VOR, and CPS approaches are available.

Cautions Helicopters and intensive flight training are in the area.

Fuel 100LL and JetA are available. American Express, Discover, MasterCard, Multi-Service, and Visa credit cards are accepted.

Frequencies
TOWER 124.8   GROUND 121.7   CLEARANCE DELIVERY 126.95
APPROACH NEW YORK 126.95   UNICOM 122.95   CTAF 124.8
ATIS 124.8.

Runway Lights Pilot-controlled lighting.

Transportation

Rental Cars Available at the terminal.
Avis, (203) 624-2161
Budget, (203) 787-1143
Hertz, (203) 777-6861
Metro, (203) 777-7777

Courtesy Cars The FBO has one available.

Area Attractions

Beinecke Rare Book Library. 121 Wall St. (203) 432-2977. This unique library houses a Gutenberg Bible and original Audubon prints plus many other interesting books and artifacts. Free

East Rock Park. Orange St. on E. Rock Rd. (203) 946-6086. Visit the Pardee Rose Gardens, bird sanctuary, hiking trails, and picnic grounds. Enjoy a superb view of the harbor and Long Island Sound. Free.

Grove Street Cemetery. Grove and Prospect Sts. Noah Webster, Eli Whitney, and many early settlers are buried here.

Lighthouse Point. End of Lighthouse Rd. Located on an 82-acre park overlooking Long Island Sound. Includes lighthouse built in 1840, antique carousel, beach, bathhouse, picnic area, and boat ramp. Admission charged.

Yale University. In the center of the city. (203) 432-2300. Founded in 1701. Visit the Yale Art Gallery, The Old Campus, Sterling Memorial and Beinecke Rare Book and Manuscript Library, Collection of Musical Instruments, Peabody Museum of Natural History and the Yale Center for British Art.

Other Attractions

Alling Memorial Golf Course. 35 Eastern St. (203) 946-8014. 18 holes, par 72. Green fees.

Historic New Haven Harbor and Long Island Sound Cruises.
Departs from Long Wharf Pier, I-95 Exit 46. (203) 562-4163. Cruise Long
Island Sound aboard the M.V. *Liberty Belle.* Moonlight and murder cruis-
es are also available.

Yale Bowl.   Chapel St. An Ivy League football field.

Interesting Facts & Events

June.   *International Festival of Arts and Ideas.* 227 Church St. (203)
498-1212.

Mid-August.   *Pilot Pen International Tennis Tournament.* Connecticut
Tennis Center. (888) 99-PILOT.

Mid-August..   *Downtown Summertime Street Festival.* (203) 946-7821.

October–May.   *Yale Repertory Theater.* 222 York St. (203) 432-1234.

For More Information   Contact the Greater New Haven Convention
and Visitor's Bureau at 1 Long Wharf, Suite 7, 06511, (203) 777-8550 or
(800) 332-STAY, for more information, a tourist package, and discounts.

Lodging

Unless otherwise noted, all lodging is within 30 minutes of the primary
airport.

The Colony Hotel
1157 Chapel St.
(203) 776-1234
Rates: $88–$108

Holiday Inn
30 Whalley Ave.
(203) 777-6221
Rates: $79–$99

Residence Inn by Marriott
3 Long Wharf Dr.
(203) 777-5337
Rates: $130–$140

Bed & Breakfast

Swan Cove
115 Sea St.
(203) 776-3240
In historic district.
Rates: $125–$175

Three Chimneys
1201 Chapel St.
(203) 789-1201
Web site:
http://www.uhranet.com/-
3chimney/
Rates: $150

Restaurants

500 Blake Street
500 Blake St.
(203) 387-0500
Specializes in Italian dishes.
Dinner: $15–$30

IndioChine Pavilion
1180 Chapel St.
(203) 865-5033
Vietnamese menu.
Dinner: $9–$18

La Mirage
111 Scrub Oak Rd.
(203) 239-1961
Specializes in stuffed shrimp.
Dinner: $11–$19

Delaware

Dover/Cheswold

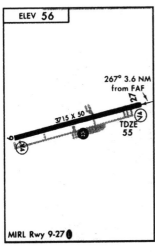

ELEV 56

267° 3.6 NM
from FAF

3715 X 50

TDZE
55

MIRL Rwy 9-27

Designed by William Penn, Dover has been the capital of Delaware since 1777. A well laid out city, you will discover many 18th- and 19th-century houses along State St. Dover also boasts having one of the biggest air cargo terminals in the world, Dover Air Force Base.

Airport Delaware Airpark (33N) is located 1 mile west of the city. Coordinates: N39° 13.10' W075° 35.79'.

Traffic Pattern The traffic pattern is flown at 856 ft MSL. Lefthand pattern for Runways 9 and 27. Runways 9 and 27 are asphalt surfaced, 3715 ft long by 50 ft wide.

FBO Diamond Aviation, Inc. (302) 674-2666. Hours: 8 A.M.–6 P.M. daily. Frequency 123.0.

Navigational Information 33N is located on the Washington Sectional Chart or L24, L28 low en route chart. From the ENO 111.4 VOR 267° at 3.8 miles.

Instrument Approaches VOR and GPS approaches are available.

Cautions Intensive flight training is in the area.

Fuel 100LL and JetA are available. Most major credit cards are accepted.

Frequencies
CLEARANCE DELIVERY 125.55 APPROACH DOVER 128.0 UNICOM/CTAF 123.0.

Runway Lights   Pilot-controlled lighting.

Transportation

Rental Cars
National, (302) 734-4446

Taxicabs
City, (302) 734-5468
Dover Limo, (302) 678-8090 or (800) 729-8091

Courtesy Cars   None available.

Area Attractions

Delaware Agricultural Museum and Village.   866 N. du Pont Hwy. (302) 734-1618. Displays of early settlement farm life to 1960. The buildings represent a late 19th-century farming community. Admission charged.

Delaware State Museum.   316 S. Governors Ave. (302) 739-4266. Three buildings devoted to exhibits of archaeology. Free.

Dover Heritage Trail.   State Visitor Center. (302) 678-2040. Guided walking tours of historic areas. Admission charged.

John Dickinson Plantation.   Near junction U.S. 113 and DE 9 on Kitts Hummock Rd. (302) 739-3277. Dickinson's restored boyhood residence. Built in 1740.

The Old State House.   Federal St. (302) 739-4266. The second oldest seat of government in continuous use in the United States. Built in 1792 and restored in 1976, the State House remains the state's symbolic capitol.

Other Attractions

Killens Pond State Park.   U.S. 13. (302) 284-4526. A 1083-acre park offering facilities for swimming, fishing, boating, hiking, and fitness trails.

Interesting Facts & Events

First weekend in May.   *Old Dover Days.* Tours of historic houses and gardens. Friends of Old Dover, (302) 674-2120.

Mid-July. *Delaware State Fair.* U.S. 13 in Harrington. View the arts and crafts and trade show, along with carnival rides and agricultural exhibits. (302) 398-3269.

September–November. *Harrington Raceway.* Fairgrounds in Harrington. Harness horse racing. (302) 398-3269.

Dover Downs. 1131 N. du Pont Hwy. (302) 674-4600. Events include NASCAR auto racing.

For More Information Contact the Kent County Delaware Convention and Visitor's Bureau at (302) 734-1736 for more information, a tourist package, and discounts.

Lodging

Unless otherwise noted, all lodging is within 30 minutes of the primary airport.

Budget Inn
1426 N. du Pont Hwy.
(302) 734-4433
Rates: $42–$55

Sheraton
1570 N. du Pont Hwy.
(302) 678-8500
Rates: $80–$90

Comfort Inn
222 S. du Pont Hwy.
(302) 674-3300
Rates: $47–$60

Restaurants

Blue Coat Inn
800 N. State St.
(302) 674-1776
Specializes in fresh seafood and Colonial recipes.

Village Inn
DE 9–Little Creek
(302) 734-3245
Specializes in stuffed flounder and prime rib.

Plaza Nine
9 E. Lockerman St.
(302) 736-9990
Specializes in fresh seafood and veal.

Wilmington

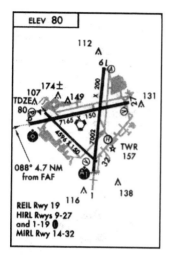

ELEV 80

Wilmington has been labeled the Chemical Capital of the World. It is also the largest city in Delaware. Settled in 1638, Wilmington's multicultural background has led to vast industrial success. The best known company is du Pont, built on Brandywine Creek in 1802.

Airport New Castle County (ILG) is located 4 miles south of the city. Coordinates: N39° 40.72' W075° 36.39'.

Traffic Pattern The traffic pattern is flown at 1080 ft MSL. Lefthand pattern for Runways 1, 9, 14, 19, 27, and 32. Runways 14 and 32 are asphalt surfaced, 4596 ft long by 150 ft wide. Runways 1 and 19 are asphalt surfaced, 7002 ft long by 200 ft wide. Runways 9 and 27 are asphalt surfaced, 7165 ft long by 150 ft wide.

FBO Atlantic Aviation. (302) 322-7350. Hours: 24. Frequency 122.95.

FBO Dawn Aeronautics, Inc. (302) 328-9695. Hours: 6 A.M.–8 P.M. daily. Frequency 122.95.

FBO Diamond Flite Center. (302) 328-1395. Hours: 8:30 A.M.–6:30 P.M. daily. Frequency 123.3.

Navigational Information ILG is located on the Washington Sectional Chart or L24, L28 low en route chart. The DQO 114.0 VOR is on the field.

Instrument Approaches ILS, VOR, NDB, VOR/DME, and GPS approaches are available.

Cautions Intensive flight training and birds are in the area.

Fuel   100LL and JetA are available. Texaco and most major credit cards are accepted.

Frequencies
TOWER   126.0   GROUND   121.7   APPROACH   PHILADELPHIA
118.35   UNICOM 122.95   CTAF 126.0   ATIS 123.95.

Runway Lights   Pilot-controlled lighting. Tower beacon dusk to dawn.

Transportation

Rental Cars   Available at the terminal.
Avis, (302) 322-2092
National, (302) 328-5636

Taxicabs
Carey Limousine Service, (303) 654-9477
Diamond, (302) 658-4321
Yellow Cab, (302) 656-8151

Courtesy Cars   The terminal and Dawn Aeronautics, Inc., have one available.

Area Attractions

Brandywine Springs Park.   4 miles W. on DE 41. (302) 323-6422. Picnicking, fireplaces, pavilions, and baseball fields for your enjoyment. Free.

Brandywine Zoo and Park.   Augustine to Market St. (302) 571-7747. The zoo features animals from North and South America. The park includes gardens, picnicking areas, and playgrounds.

Grand Opera House.   818 Market St. Mall. (302) 652-5577. Historic landmark built in 1871 by Masons. It now serves as Delaware's Center for the Performing Arts and Grand Opera.

Winterthur Museum, Garden, and Library.   6 miles N.W. on DE 52. (302) 888-4600 or (800) 448-3883. Decorative arts and American antiques representing the period from 1640 to 1860 are displayed. Admission charged.

Other Attractions

Amtrak Station. Martin Luther King Blvd. and French St. Victorian railroad station still in use.

Banning Park. Middleboro Rd. and Maryland Ave. (302) 323-6422. Areas include picnicking, fishing, tennis courts, and ball fields. Free.

Interesting Facts & Events

Last Saturday in April. *Hagley's Irish Festival.* Irish dancers and singers. (302) 658-2400.

Delaware Nature Society Harvest Moon Festival. Ashland Nature Center. Cider-pressing demonstrations, hayrides, arts and crafts, and musical entertainment. (302) 239-2334.

Delaware Park. I-95, Exit 4B. Thoroughbred racing. (302) 994-2521.

For More Information Contact the Greater Wilmington Convention and Visitor's Bureau at (302) 652-4088 for more information, a tourist package, and discounts.

Lodging

Unless otherwise noted, all lodging is within 30 minutes of the primary airport.

Courtyard by Marriott
1102 West St.
(302) 429-7600
Rates: $99–$130

Holiday Inn–Downtown
700 King St.
(302) 655-0400
Rates: $65–$125

Hotel Du Pont
11th and Market Sts.
(302) 594-3100
Rates: $139–$495

Sheraton Suites
422 Delaware Ave.
(302) 654-8300
Rates: $165–$195

Bed & Breakfast

The Boulevard Bed and Breakfast
1909 Baynard Blvd.
(302) 656-9700
Rates: $55–$85

Darley Manor
3701 Philadelphia Pike
(302) 792-2127
Rates: $79–$99

Restaurants

Constantinou's Beef and Seafood
1616 Delaware Ave.
(302) 652-0653
Specializes in steak and seafood.

India Palace
101 N. Maryland Ave.
(302) 655-8772
Specializes in vegetarian and
tandoori dishes.

Picciotti's
3001 Lancaster Ave.
(302) 652-3563
Specializes in filet mignon.

Waterworks Café
16th and French Sts.
(302) 652-6022
Specializes in fresh seafood.

Illinois

Chicago

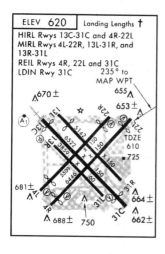

ELEV 620	Landing Lengths †

HIRL Rwys 13C-31C and 4R-22L
MIRL Rwys 4L-22R, 13L-31R, and 13R-31L
REIL Rwys 4R, 22L and 31C
LDIN Rwy 31C   235° to MAP WPT

670±   655
653±
TDZE 610
725
681±   664±
688± 750   662±

Nicknamed the Windy City, Chicago has the image of a working-class town, yet it has more museums, zoos, planetariums, and other cultural activities than many larger eastern cities.

Airport Midway (MDW) is located 9 miles south of the city. Coordinates: N41° 47.16' W087° 45.15'.

Traffic Pattern The traffic pattern is flown at 1419 ft MSL. Lefthand pattern for Runways 4L, 4R, 13L, 13C, 13R, 22L, 22R, 31L, 31C, and 31R. Runways 13C and 31C are concrete surfaced, 6522 ft long by 150 ft wide. Runways 13L and 31R are asphalt surfaced, 5142 ft long by 150 ft wide. Runways 13R and 31L are concrete surfaced, 3859 ft long by 60 ft wide. Runways 4L and 22R are asphalt surfaced, 5509 ft long by 150 ft wide. Runways 4R and 22L are concrete surfaced, 6446 ft long by 150 ft wide.

FBO Aero Services International, Inc. (312) 582-5720. Hours: 24. Frequency 122.95.

FBO Million Air Midway. (312) 284-2867. Hours: 24. Frequency 122.95.

FBO Monarch Air Service, Inc. (312) 471-4530. Hours: 24. Frequency 122.95.

FBO Signature Flight Support. (312) 767-4400. Hours: 24. Frequency 122.95.

Navigational Information MDW is located on the Chicago Sectional Chart or L23 low en route chart. From the CGT 114.2 VOR 332° at 18.5 miles.

Instrument Approaches ILS, NDB, MLS, VOR/DME, and GPS approaches are available.

Cautions Buildings, chimneys, and birds are in the vicinity.

Fuel 100LL and JetA are available. Exxon, Texaco, Amoco, Phillips, and most major credit cards are accepted.

Frequencies
TOWER 118.7   GROUND 121.7   CLEARANCE DELIVERY 121.85
APPROACH CHICAGO 118.4   CHICAGO 126.05   UNICOM 122.95
ATIS 132.75.

Runway Lights Operated by the tower.

Transportation

Rental Cars
Alamo, (312) 581-4531
Avis, (312) 284-3640
Budget, (312) 686-6769
Dollar, (312) 735-7200
Hertz, (312) 735-7272
National, (312) 471-3450
Rent-a-Wreck, (312) 585-7344 or (800) 458-2899

Taxicabs
Airtran Limo, (800) 972-0500
C&W Transport, (312) 493-2700
Carey Limousine Service, (312) 763-0009
Checker and Yellow Cabs, (312) 829-4222
Continental Express, (312) 454-7800
Tri-State, (800) 248-8747

Courtesy Cars None available.

Chicago/ Prospect Heights/ Wheeling

Airport   Palwaukee Municipal (PWK) is located 18 miles northwest of the city. Coordinates: N42° 06.82' W087° 54.09'.

Traffic Pattern   The traffic pattern is flown at 1447 ft MSL. Lefthand pattern for Runways 6, 12L, 12R, 16, 24, 30L, 30R, and 34. Runways 12L and 30R are asphalt surfaced, 4397 ft long by 50 ft wide. Runways 12R and 30L are 3228 ft long by 40 ft wide. Runways 16 and 34 are asphalt surfaced, 5137 ft long by 100 ft wide. Runways 6 and 24 are asphalt surfaced, 3652 ft long by 50 ft wide.

FBO   Priester Aviation. (847) 537-1200. Hours: 24. Frequency 122.95.

FBO   Service Aviation. (847) 808-9690. Hours: 5 A.M.–11 P.M. daily. Frequency 122.95.

Navigational Information   PWK is located on the Chicago Sectional Chart or L23 low en route chart. From the OBK 113.0 VOR 159° at 7 miles.

Instrument Approaches   ILS, VOR, and GPS approaches are available.

Cautions   Construction, power lines, and birds are in the vicinity.

Fuel   100LL and JetA are available. Phillips, Avfuel, and most major credit cards are accepted.

Frequencies
TOWER 119.9, 124.7   GROUND 121.7   CLEARANCE DELIVERY 124.7   APPROACH CHICAGO 120.55   UNICOM 122.95   CTAF 119.9   ATIS 124.2.

Runway Lights   Tower and pilot-controlled lighting.

Transportation

Rental Cars
Budget, (847) 465-2080
Hertz, (847) 465-1156
Snappy, (847) 537-6200

Taxicabs
Bob's Cab Co., (847) 729-0303
Carey Limousine Service, (847) 763-0009
City Wide Limo, (847) 537-6200

Courtesy Cars Priester Aviation and Service Aviation have one available.

Chicago/West Chicago

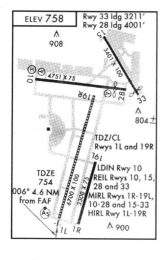

ELEV 758
Rwy 33 ldg 3211'
Rwy 28 ldg 4001'
908
4751 X 75
3401 X 100
804
TDZ/CL
Rwys 1L and 19R
TDZE 754
006° 4.6 NM from FAF
LDIN Rwy 10
REIL Rwys 10, 15, 28 and 33
MIRL Rwys 1R-19L, 10-28 and 15-33
HIRL Rwy 1L-19R
900

Airport DuPage (DPA) is located 29 miles west of the city. Coordinates: N41° 54.40' W088° 14.91'.

Traffic Pattern The traffic pattern is flown at 1557 ft MSL. Lefthand pattern for Runways 1L, 1R, 10, 15, 19L, 19R, 28, and 33. Runways 1L and 19R are concrete surfaced, 5100 ft long by 100 ft wide. Runways 1R and 19L are concrete surfaced, 3300 ft long by 75 ft wide. Runways 10 and 28 are asphalt surfaced, 4751 ft long by 75 ft wide. Runways 15 and 33 are asphalt surfaced, 3401 ft long by 100 ft wide.

FBO DuPage Flight Center. (630) 208-5600. Hours: 5 A.M.–11 P.M. Frequency 122.95.

Navigational Information DPA is located on the Chicago Sectional Chart or L23 low en route chart. From the DPA 108.4 VOR 076° at 4.7 miles.

Instrument Approaches ILS, VOR, and GPS approaches are available.

Fuel 100LL and JetA are available. Phillips and most major credit cards are accepted.

Frequencies
TOWER 120.9, 124.5   GROUND 121.8   CLEARANCE DELIVERY 119.75   APPROACH CHICAGO 133.5   UNICOM 122.95   ATIS 124.8.

Runway Lights Operated by the tower.

Transportation

Rental Cars

Budget, (630) 513-3000

Enterprise, (630) 377-7800

U.S. Car and Truck, (630) 377-1234

Taxicabs

A-1 Limo, (630) 858-3317

American Cab, (630) 232-1363

Dukane, (630) 293-1171

Fox Cab, (630) 232-2227

J.I.L. Limo, (630) 888-1344

Limo Network West, (630) 584-1616

Courtesy Cars   The FBO has one available.

Area Attractions

Chicago Board of Trade.   141 W. Jackson Blvd. (312) 435-3590. Commodity futures exchange. Free.

Chicago Mercantile Exchange.   30 S. Wacker Dr. (312) 930-8249. Leading financial futures exchange. Free.

Lincoln Park Zoological Gardens.   Webster Ave. and Stockton Dr. or Cannon Dr. off Fullerton Ave. (312) 742-2000. The zoo houses more than 1600 animals including a gorilla collection. Free.

Other Attractions

Balmoral Park Race Track.   312) 568-5700. Harness racing.

Chicago Bears.   Soldier Field, 1600 S. Waldron. (847) 295-6600. Professional football.

Chicago Bulls.   United Center, 1901 W. Madison St. (312) 455-4000. Professional basketball.

Chicago Cubs.   Wrigley Field, 1060 W. Addison St. (773) 404-2827. Major league baseball.

Chicago White Sox. Comiskey Park, 333 W. 35th St. (312) 674-1000. Major league baseball.

Interesting Facts & Events

Second weekend in February. *Chicago Auto Show.* Foreign and domestic cars are displayed.

Weekend closest to March 17. *St. Patrick's Day Parade.*

Late August. *Air and Water Show.* Blue Angels and Golden Knights parachute teams.

Three weeks in October. *Chicago International Film Festival.* (312) 644-3400.

For More Information Contact the Chicago Office of Tourism, Chicago Cultural Center at 78 E. Washington St. (312) 744-2400 or (800) 226-6632, for more information, a tourist package, and discounts.

Lodging

Unless otherwise noted, all lodging is within 30 minutes of the primary airport.

Four Seasons
120 E. Delaware Pl., north of
the Loop
(312) 280-8800
e-mail: http://www.fourtiff@aol.com
Rates: $325–$995

Hampton Inn
6540 S. Cicero Ave., south of
Midway Airport
(708) 496-1900
Free airport transportation.
Rates: $89–$99

Hilton and Towers
720 S. Michigan Towers, south of
the Loop
(312) 922-4400
Rates: $165–$250

Holiday Inn–Midway Airport
7353 S. Cicero Ave.
(773) 581-5300
Rates: $84–$94

Hyatt at University Village
625 S. Ashland Ave., west of
the Loop
(312) 243-7200
Rates: $120–$560

Ramada Congress
520 S. Michigan Ave., south of
the Loop
(312) 427-3800
Rates: $75–$700

Renaissance
1 W. Wacker Dr., on the
Chicago River
(312) 372-7200
Rates: $270–$900

The Ritz–Carlton Hotel
160 E. Pearson St. north of
the Loop
(312) 266-1000
Rates: $315–$1050

Sleep Inn
6650 S. Cicero Ave., near
Midway Airport
(708) 594-0001
Free airport transportation.
Rates: $79–$85

Bed & Breakfast

Gold Coast Guest House
113 W. Elm St.
(312) 337-0361
e-mail:
http://www.bbchicago.com
Renovated brick townhome built
in 1873.
Rates: $119–$165

Restaurants

Arun's
4156 N. Kedzie Ave., north of
the Loop
(773) 539-1909
Specializes in three-flavored
red snapper.
Dinner: $12–$25

Everest
440 S. La Salle St.
40th floor of Midwest
Stock Exchange
(312) 663-8920
Specializes in seafood.
Dinner: $23–$35

Michael Jordan's
500 N. La Salle St.
(312) 644-3865
Specializes in Michael's "Nothin'
but Net" burger.
Michael Jordan memorabilia
displayed.
Dinner: $12–$27

Ritz–Carlton Dining Room
The Ritz–Carlton Hotel
(312) 227-5866
Specializes in New Zealand
venison steak.
Dinner: $26–$37

Walter's
28 Main, 3.5 miles N.E. of
O'Hare Airport
(847) 825-2240
Specializes in grilled seafood.
Dinner: $5–$27

Peoria

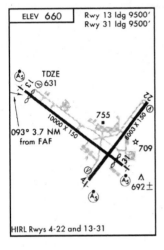

Peoria is in the heart of a rich agricultural area on the Illinois River. It is the oldest settlement in the state.

Airport Greater Peoria Regional (PIA) is located 4 miles west of the city. Coordinates: N40° 39.85' W089° 41.60'.

Traffic Pattern The traffic pattern is flown at 1700 ft MSL. Lefthand pattern for Runways 4, 13, 22, and 31. Runways 13 and 31 are asphalt surfaced, 10,000 ft long by 150 ft wide. Runways 4 and 22 are asphalt surfaced, 8008 ft long by 150 ft wide.

FBO Byerly Aviation. (309) 697-6300. Hours: 24. Frequency 122.95.

Navigational Information PIA is located on the Chicago Sectional Chart or L11, L23 low en route chart. From the PIA 115.2 VOR 098° at 4.6 miles.

Instrument Approaches ILS, VOR, VOR/DME, VORTAC, NDB, ASR, and GPS approaches are available.

Fuel 100LL and JetA are available. Avfuel and most major credit cards are accepted.

Frequencies
TOWER 119.1 GROUND 121.6 CLEARANCE DELIVERY 121.85
APPROACH PEORIA 125.8, PEORIA 119.95 UNICOM 122.95
ATIS 126.10.

Runway Lights Operated by the tower from dusk until dawn and IFR.

Transportation

Rental Cars
Avis, (309) 697-5200
Budget, (309) 697-2722
Hertz, (309) 697-0650
National, (309) 697-0566
S & K, (309) 697-8267

Taxicabs
Community Cab, (309) 676-0064
Pearlene Bell Cab, (309) 674-5956
Yellow Checker, (309) 676-0731

Courtesy Cars   None available.

Peoria

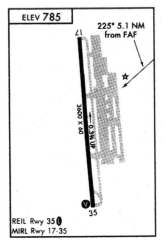

ELEV **785**

225° 5.1 NM
from FAF

3600 X 60

0.3% UP

35

REIL Rwy 35
MIRL Rwy 17-35

Airport Mount Hawley Auxiliary (3MY) is located 7 miles north of the city. Coordinates: N40° 47.72' W089° 36.80'.

Traffic Pattern The traffic pattern is flown at 1775 ft MSL. Lefthand pattern for Runways 17 and 35. Runways 17 and 35 are asphalt surfaced, 600 ft long by 60 ft wide.

FBO North Point Aviation, Inc. (309) 693-1908. Hours: daylight. Frequency 122.7.

Navigational Information 3MY is located on the Chicago Sectional Chart or L11, L23 low en route chart. From the PIA 115.2 VOR 046° at 10.7 miles.

Instrument Approaches A VOR approach is available.

Fuel 100LL and JetA are available. Self-service pump. Most major credit cards are accepted.

Frequencies
APPROACH PEORIA 125.8   UNICOM/CTAF 122.7.

Runway Lights Operated by the tower.

Transportation

Rental Cars
Avis, (309) 673-3081

Taxicabs
Community Cab, (309) 676-0064
Pearlene Bell Cab, (309) 674-5956
Yellow Checker, (309) 676-0731

Courtesy Cars The terminal has one available.

Area Attractions

Eureka College. U.S. 24. (309) 467-6318. One of its most famous graduates is Ronald Reagan. See a collection of Reagan memorabilia.

Glen Oak Park and Zoo. Prospect Rd. and McClure Ave. (309) 682-1200. More than 250 species. Admission charged.

Lakeview Museum of Arts and Sciences. 1125 W. Lake Ave. (309) 686-7000. Exhibits in the arts and sciences. Children's Discovery Center. Admission charged.

Other Attractions

Spirit of Peoria. Departs from The Landing at the foot of Main St. (309) 699-7232 or (800) 676-8988. Replica of a turn-of-the-century stern-wheeler. 1.5 hour sightseeing cruise. Admission charged.

Wheels o' Time Museum. 11923 N. Knoxville Ave. (309) 243-9020. Antique autos, tractors, fire engines, and railroad memorabilia. Admission charged.

Interesting Facts & Events

Four days at Father's Day weekend. *Steamboat Days.* Riverboat races.

For More Information Contact the Convention and Visitor's Bureau at 403 N. E. Jefferson, (309) 676-0303 or (800) 747-0302, for more information, a tourist package, and discounts.

Lodging

Unless otherwise noted, all lodging is within 30 minutes of the primary airport.

Best Western Eastlight Inn
401 N. Main St.
(309) 699-7231
Free airport transportation.
Rates: $56–$64

Fairfield Inn by Marriott
4203 N. War Memorial Dr.
(309) 686-7600
Rates: $45–$65

Hampton Inn
11 Winners Way
(309) 694-0711
Free airport transportation.
Rates: $69–$150

Signature Inn
4112 N. Brandywine
(309) 685-2556
Rates: $60–$70

Restaurants

Jumer's
117 N. Western Ave.
(309) 673-8181
Specializes in Wiener schnitzel
and pork roast.
Dinner: $8–$15

Stephanie's
1825 N. Knoxville Ave.
(309) 682-7300
Specializes in midwestern
American cooking.
Dinner: $10–$20

Indiana

Evansville

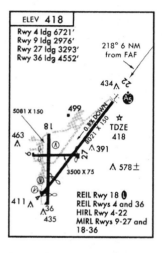

```
ELEV 418
Rwy 4 ldg 6721'
Rwy 9 ldg 2976'
Rwy 27 ldg 3293'       218° 6 NM
Rwy 36 ldg 4552'       from FAF

                    434 ∧

5081 X 150      499
                          ☆
       8 L             TDZE
463                    418
∧      ∧           ∧ 391

                    ∧ 578±
       3500 X 75

             REIL Rwy 18
411 ∧        REIL Rwys 4 and 36
    ∧ 36     HIRL Rwy 4-22
    435      MIRL Rwys 9-27 and
             18-36
```

Located on the Ohio River, Evansville is the largest city in southern Indiana. The Ohio River offers visitors plenty of recreational activities.

Airport Evansville Regional (EVV) is located 3 miles north of the city. Coordinates: N38° 02.28' W087° 31.84'.

Traffic Pattern The traffic pattern is flown at 1500 ft MSL. Lefthand pattern for Runways 4, 9, 18, 22, 27, and 36. Runways 18 and 36 are asphalt surfaced, 5081 ft long by 150 ft wide. Runways 4 and 22 are asphalt surfaced, 8021 ft long by 150 ft wide. Runways 9 and 27 are asphalt surfaced, 3500 ft long by 100 ft wide.

FBO Million Air Evansville. (812) 425-4700. Hours: 24. Frequency 122.95.

FBO Tri-State Aero, Inc. (812) 426-1221. Hours: 24. Frequency 122.95.

Navigational Information EVV is located on the St. Louis Sectional Chart or L21 low en route chart. From the PXV 113.3 VOR 057° at 12.8 miles.

Instrument Approaches ILS, VOR, NDB, ASR, and GPS approaches are available.

Fuel 100LL and JetA are available. Exxon, Phillips, and most major credits cards are accepted.

Frequencies
TOWER 118.7   GROUND 121.9   CLEARANCE DELIVERY 126.6
APPROACH EVANSVILLE 127.35, INDIANAPOLIS CENTER 128.3
UNICOM 122.95   CTAF 118.7   ATIS 120.2.

Runway Lights   Operated by the tower dusk until dawn.

Transportation

Rental Cars   Located at the terminal.
Avis, (812) 423-5645
Budget, (812) 423-4343
Hertz, (812) 425-7143
National, (812) 425-2426

Taxicabs
Bassemiers Transportation, (812) 477-8000
River City Taxi, (812) 429-0000

Courtesy Cars   Million Air Evansville has one available.

Area Attractions

Evansville Museum of Arts and Science.   411 S. E. Riverside Dr.
(812) 425-2406. History and science exhibits. Loch Planetarium (admission charged) and steam train. Free.

Mesker Park Zoo.   N.W. edge of town in Mesker Park. (812) 428-0715. More than 700 animals including a children's zoo, tour train, and paddleboats. Admission charged.

Other Attractions

Angel Mounds State Historic Site.   8215 Pollack Ave. (812) 853-3956. Large group of prehistoric mounds. Exhibits and artifacts. Free.

Burdette Park.   (812) 435-5602. County park with fishing, picnicking, waterslides, miniature golf, and more. Some fees charged.

Interesting Facts & Events

Mother's Day weekend. *Ohio River Arts Festival.*

June 14–July 4. *Freedom Festival.* Citywide events including parade, fireworks, food, and more. (812) 464-9576.

For More Information Contact the Evansville Convention and Visitor's Bureau at 401 S.E. Riverside Dr., (812) 425-5402 or (800) 433-3025, for more information, a tourist package, and discounts.

Lodging

Unless otherwise noted, all lodging is within 30 minutes of the primary airport.

Days Inn
5701 U.S. 41 N.,
near Regional Airport
(812) 464-1010
Free airport transportation.
Rates: $54–$125

Fairfield Inn by Marriott
7879 Eagle Crest Blvd.
(812) 471-7000
Rates: $55–$66

Holiday Inn–Airport
4101 U.S. 41 N.,
near Regional Airport
(812) 424-6400
Free airport transportation.
Rates: $79–$89

Marriott–Airport
7101 U.S. 41 N., adjacent to
Regional Airport
(812) 867-7999
Free airport transportation.
Rates: $99–$350

Restaurants

Elliott's
4701 E. Powell Ave.
(812) 473-3378
Specializes in steak and seafood.
Dinner: $8–$20

Fort Wayne

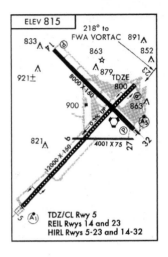

ELEV 815
218° to
FWA VORTAC 891
833
863 852
921±
879 TDZE
8000 X 150 800
900 863
821 4001 X 75

TDZ/CL Rwy 5
REIL Rwys 14 and 23
HIRL Rwys 5-23 and 14-32

First settled by fur traders, the city's location on a river junction has turned Fort Wayne into the second largest city in Indiana.

Airport Fort Wayne International (FWA) is located 7 miles southwest of the city. Coordinates: N40° 58.70′ W085° 11.69′.

Traffic Pattern The traffic pattern is flown at 1801 ft MSL. Lefthand pattern for Runways 9, 14, 15, 23, 27, and 32. Runways 14 and 32 are asphalt surfaced, 8500 ft long by 150 ft wide. Runways 15 and 23 are asphalt surfaced, 12,000 ft long by 150 ft wide. Runways 9 and 27 are asphalt and concrete surfaced, 4001 ft long by 75 ft wide.

FBO Consolidated Airways. (219) 747-4189. Hours: 6 A.M.–10 P.M. weekdays only, 24 hours on request. Frequency 123.3.

FBO Fort Wayne Air Service. (219) 747-1565. Hours: 6 A.M.–midnight daily, after on request. Frequency 122.95.

Navigational Information FWA is located on the Chicago Sectional Chart or L23 low en route chart. The FWA 117.8 VOR is on the field.

Instrument Approaches ILS, VOR, VORTAC, NDB, ASR, and GPS approaches are available.

Fuel 100LL and JetA are available. Phillips, Air BP, and most major credit cards are accepted.

Frequencies
TOWER 119.1   GROUND 121.9   CLEARANCE DELIVERY 124.75
APPROACH FORT WAYNE 132.15SE/NE, FORT WAYNE 127.2SW/NW,
FORT WAYNE 135.325   UNICOM 122.95   ATIS 121.25.

Runway Lights Operated by the tower.

Transportation

Rental Cars

Avis, (219) 747-7438

Budget, (219) 448-1696

Hertz, (219) 747-6108

National, (219) 747-4124

Taxicabs

Checker, (219) 426-8555

Deluxe, (219) 482-3634

Courtesy Cars The terminal has one available.

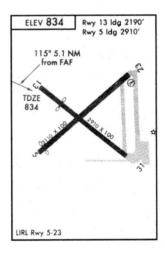

Airport Smith Field (SMD) is located 4 miles north of the city. Coordinates: N41° 08.60' W085° 09.17'.

Traffic Pattern The traffic pattern is flown at 1633 ft MSL. Lefthand pattern for Runways 5, 13, 23, and 31. Runways 13 and 31 are asphalt surfaced, 2910 ft long by 100 ft wide. Runways 5 and 23 are asphalt surfaced, 3110 ft long by 100 ft wide.

FBO Bowman Aviation, Inc. (219) 489-5517. Hours: 7 A.M.–dusk. Frequency 122.8.

Navigational Information SMD is located on the Chicago Sectional Chart or L23 low en route chart. From the FWA 117.8 VOR 009° at 10 miles.

Instrument Approaches   VOR and GPS approaches are available.

Fuel   100LL and JetA are available. Phillips and most major credit cards are accepted.

Frequencies
CLEARANCE DELIVERY 126.6   APPROACH FORT WAYNE 127.2, FORT WAYNE 132.15   UNICOM/CTAF 122.8.

Runway Lights   Operated by the tower.

Transportation

Rental Cars
Ace, (219) 482-8546
Enterprise, (219) 482-2662

Taxicabs
Available at the terminal.

Courtesy Cars   None available.

Area Attractions

Fort Wayne Children's Zoo.   3411 Sherman Blvd. in Franke Park. (219) 427-6800. Exotic animals, pony rides, train rides, and more. Admission charged.

Fort Wayne Museum of Art.   311 E. Main St. (219) 422-6467. Exhibitions, art classes, interactive programs, and lectures. Admission charged.

The New Lincoln Museum.   200 E. Berry St. (219) 455-3864. Literature, paintings, and a number of Lincoln's personal items. Admission charged.

Other Attractions

Foellinger–Freimann Botanical Conservatory.   1100 S. Calhoun St. (219) 427-6440. Seasonally changing displays of colorful flowers, exotic plants, cascading waterfalls, and more. Admission charged.

Lakeside Rose Garden.   1500 Lake Ave. (219) 427-1267. Display rose garden. Free.

Interesting Facts & Events

Mid-June, 4 days.   *Germanfest.* Celebration of the city's German heritage. (800) 767-7752.

Mid-July, 9 days.   *Three Rivers Festival.* Parades, balloon races, crafts, games, music, fireworks, and more. (219) 745-FEST.

For More Information   Contact the Fort Wayne/Allen County Convention and Visitor's Bureau at 1021 S. Calhoun St., (219) 424-3700 or (800) 767-7752, for more information, a tourist package, and discounts.

Lodging

Unless otherwise noted, all lodging is within 30 minutes of the primary airport.

Courtyard by Marriott
1619 W. Washington Center Rd.
(219) 489-3273
Rates: $69–$229

Hilton
1020 S. Calhoun St.
(219) 420-1100
Free airport transportation.
Rates: $89–$350

Holiday Inn
3939 Ferguson Rd.,
near Fort Wayne airport
(219) 747-9171
Free airport transportation.
Rates: $78–$150

Holiday Inn–Downtown
300 E. Washington Blvd.
(219) 422-5511
Free airport transportation.
Rates: $89–$190

Restaurants

Café Johnell
2529 S. Calhoun St.
(219) 456-1939
Specializes in caneton a l'orang
flambé.
Dinner: $15–30

Elegant Farmer
1820 Coliseum Blvd. N.
(219) 482-1976
Specializes in steak and seafood.
Dinner: $7–$14

Indianapolis

ELEV 823
207° 6 NM from FAF
TDZE 820
MIRL Rwy 3-21
REIL Rwy 3

Indianapolis, the state capital, is the state's largest city. Home of the Indy 500, it is also known as the nation's amateur sports capital.

Airport Eagle Creek Airpark (EYE) is located 7 miles west of the city. Coordinates: N39° 49.84' W086° 17.66'.

Traffic Pattern The traffic pattern is flown at 1600 ft MSL. Lefthand pattern for Runways 3 and 21. Runways 3 and 21 are asphalt surfaced, 4200 ft long by 75 ft wide.

FBO Eagle Creek Aviation Services, Inc. (317) 293-6935. Hours: 7 A.M.–9 P.M. daily. Frequency 122.8.

FBO Indiana Aircraft Sales, Inc. (317) 293-4548. Hours: 8:30 A.M.–6 P.M. weekdays, weekends by appointment. Frequency 122.95.

Navigational Information EYE is located on the St. Louis Sectional Chart or L23 low en route chart. From the VHP 116.3 VOR 073° at 3.5 miles.

Instrument Approaches LOC, VOR, NDB, and GPS approaches are available.

Fuel 100LL and JetA are available. Texaco and most major credit cards are accepted.

Frequencies
CLEARANCE DELIVERY 128.6   DEPARTURE 119.05   APPROACH INDIANAPOLIS 124.95, INDIANAPOLIS 119.05   UNICOM/CTAF 122.8.

Runway Lights Pilot-controlled lighting.

Transportation

Rental Cars

Enterprise, (317) 299-0711

Taxicabs

Yellow, (317) 637-5421

Courtesy Cars The terminal has one available.

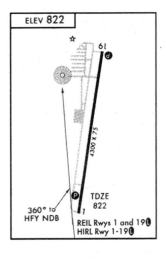

Airport Greenwood Municipal (HFY) is located 3 miles northeast of the city. Coordinates: N39° 37.71' W086° 05.27'.

Traffic Pattern The traffic pattern is flown at 1600 ft MSL. Lefthand pattern for Runways 1 and 19. Runways 1 and 19 are asphalt surfaced, 4300 ft long by 75 ft wide.

FBO Greenwood Aviation. (317) 881-0887. Hours: 8 A.M.–8 P.M.

Navigational Information HFY is located on the St. Louis Sectional Chart or L21, L23 low en route chart. From the SHB 112.0 VOR 268° at 12.2 miles.

Instrument Approaches VOR and NDB approaches are available.

Fuel 100LL and JetA are available. Most major credit cards are accepted.

Frequencies
DEPARTURE 124.95   APPROACH INDIANAPOLIS 124.95   UNICOM/
CTAF 123.0.

Runway Lights   Tower and pilot-controlled lighting.

Transportation

Rental Cars
Enterprise, (317) 881-3774
National, (317) 885-1530

Taxicabs
Greenwood, (317) 881-2525

Courtesy Cars   None available.

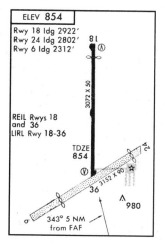

Airport   Indianapolis   Brookside Airpark (I21) is located 1 mile north of the city. Coordinates: N39° 54.29′ W085° 55.45′.

Traffic Pattern   The traffic pattern is flown at 1654 ft MSL. Lefthand pattern for Runways 18 and 36. Runways 18 and 36 are asphalt surfaced, 3072 ft long by 50 ft wide.

FBO   Brookside Airpark. (317) 335-2090. Hours: 8 A.M.–5 P.M.

Navigational Information   I21 is located on the St. Louis Sectional Chart or L23 low en route chart. From the SHB 112.0 343° at 17 miles.

Instrument Approaches   VOR and GPS approaches are available.

Fuel   100LL is available.

Frequencies
APPROACH INDIANAPOLIS 127.15   UNICOM/CTAF 122.8.

Runway Lights   Operated by the tower.

Transportation

Rental Cars   Available at the terminal.

Taxicabs
Available at the terminal.

Courtesy Cars   Available at FBO.

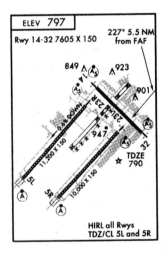

Airport Indianapolis International (IND) is located 7 miles southwest of the city. Coordinates: N39° 43.04' W086° 17.66'.

Traffic Pattern The traffic pattern is flown at 1597 ft MSL. Lefthand pattern for Runways 5L, 5R, 14, 23L, 23R, and 32. Runways 5L and 23R are asphalt surfaced, 11,200 ft long by 200 ft wide. Runways 5R and 23L are concrete surfaced, 10,000 ft long by 150 ft wide. Runways 14 and 32 are asphalt surfaced, 7604 ft long by 150 ft wide.

FBO AMR   COMBS–Indianapolis, Inc. (317) 248-4900. Hours: 24. Frequency 122.95.

FBO Raytheon Aircraft Services. (800) 365-6734. Hours: 24. Frequency 122.95.

Navigational Information IND is located on the St. Louis Sectional Chart or L21, L23 low en route chart. From the VHP 116.3 VOR 146° at 6.8 miles.

Instrument Approaches ILS, VOR, NDB, and ASR approaches are available.

Cautions Birds are reported in the vicinity.

Fuel 100LL and JetA are available. Texaco, Phillips, and most major credit cards are accepted.

Frequencies
TOWER 120.9, 123.95   GROUND 121.9, 121.8   CLEARANCE DELIVERY 128.75   DEPARTURE 119.5   APPROACH INDIANAPOLIS 127.15, INDIANAPOLIS 124.65   UNICOM 122.95   ATIS 124.4.

Runway Lights Operated by the tower.

Transportation

Rental Cars
Alamo, (800) 377-9633
Avis, (800) 331-1212
Budget, (800) 527-0700
Hertz, (800) 654-3131
National, (317) 243-1150
Thrifty, (800) 367-2277

Taxicabs
Available at the terminal.

Courtesy Cars AMR COMBS–Indianapolis, Inc., has one available.

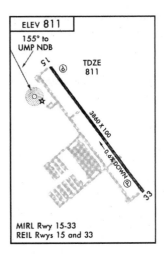

Airport Indianapolis Metropolitan (UMP) is located 8 miles northeast of the city. Coordinates: N39° 56.11' W086° 02.70'.

Traffic Pattern The traffic pattern is flown at 1600 ft MSL. Lefthand pattern for Runways 15 and 33. Runways 15 and 33 are asphalt surfaced, 3860 ft long by 100 ft wide.

FBO Indianapolis Aviation, Inc. (317) 849-0840. Hours: 7 A.M.–9 P.M. Monday–Saturday; 8 A.M.–9 P.M. Sunday. Frequency 123.0.

Navigational Information UMP is located on the St. Louis Sectional Chart or L23 low en route chart. From the SHB 112.0 VOR 330° at 20.8 miles.

Instrument Approaches VOR, NDB, and GPS approaches are available.

Fuel 100LL and JetA are available. Avfuel and most major credit cards are accepted.

Frequencies
APPROACH INDIANAPOLIS 127.15   UNICOM/CTAF 123.0.

Runway Lights Operated by the tower.

Transportation

Rental Cars
Enterprise, (317) 845-2923
National, (317) 833-9011

Taxicabs
Carmel Yellow Cab, (317) 844-4244
Noblesville Yellow Cab, (317) 773-5412

Courtesy Cars None available.

Airport Indianapolis Terry (I52) is located 14 miles northwest of the city. Coordinates: N40° 01.84' W086° 15.09'.

Traffic Pattern The traffic pattern is flown at 1923 ft MSL. Lefthand pattern for Runways 18 and 36. Runways 18 and 36 are asphalt surfaced, 5500 ft long by 100 ft wide.

FBO Indy Flight Center, Inc. (317) 873-5563. Hours: 24. Frequency 122.95.

FBO Montgomery Aviation, Inc. (317) 769-4487. Hours: 8 A.M.–5 P.M. weekdays. Frequency 122.95.

FBO PAD, Inc. (317) 769-6969. Hours: 8 A.M.–6 P.M. Frequency 122.95.

FBO Van S Aviation Corp. (317) 873-5563, (317) 769-3288, or (800) 968-3779. Hours: 8 A.M.–6 P.M. Monday–Saturday, 9 A.M.–6 P.M. Sunday. Frequency 123.05.

Navigational Information I52 is located on the Chicago Sectional Chart or L23 low en route chart. From the VHP 116.3 VOR 021° at 14 miles.

Instrument Approaches ILS, VOR, NDB, VOR/DME, and GPS approaches are available.

Fuel 100LL and JetA are available. Phillips and most major credit cards are accepted.

Frequencies
APPROACH INDIANAPOLIS 127.15   UNICOM/CTAF 123.05.

Runway Lights Tower and pilot-controlled lighting.

Transportation

Rental Cars
Amtram, (317) 577-3100
Enterprise, (317) 848-9344 or (800) 325-8007 (ask for Carmel office)
Hertz, (317) 243-5234
National, (317) 844-9011

Taxicabs
Yellow Cab, (317) 873-3664

Courtesy Cars   None available.

ELEV 862

TDZE
861

252° 5.4 NM
from FAF

5500 X 100

3901 X 75

916

REIL Rwys 7 and 34
HIRL Rwy 7-25 ◗
MIRL Rwy 16-34

Airport   Mount Comfort (Indianapolis) (MQJ) is located 12 miles east of the city. Coordinates: N39° 50.61' W085° 53.82'.

Traffic Pattern   The traffic pattern is flown at 1800 ft MSL. Lefthand pattern for Runways 7, 16, 25, and 34. Runways 16 and 34 are concrete surfaced, 3901 ft long by 75 ft wide. Runways 7 and 25 are concrete surfaced, 5500 ft long by 100 ft wide.

FBO   Angel Air, Inc. (317) 335-3371. Hours: 7 A.M.–8 P.M. daily. Frequency 122.7.

Navigational Information   MQJ is located on the St. Louis Sectional Chart or L23 low en route chart. From the SHB 112.0 VOR 344° at 13.1 miles.

Instrument Approaches   ILS, VOR, and GPS approaches are available.

Fuel   100LL and JetA are available. Avfuel and most major credit cards are accepted.

Frequencies
CLEARANCE DELIVERY 119.25   APPROACH INDIANAPOLIS 127.15
UNICOM/CTAF 122.7.

Runway Lights   Tower and pilot-controlled lighting.

Transportation

Rental Cars
National, (317) 335-3371

Taxicabs
Lawrence Rainbow, (317) 897-2122
Metro Cab, (317) 634-1112
Yellow Cab, (317) 637-5421

Courtesy Cars   The terminal has one available.

Area Attractions

Conner Prairie.   13400 Allisonville Rd. (317) 776-6000 or (800) 966-1836. A 250-acre historic site with costumed interpreters who depict the life and times of a settlement in the 1800s. Admission charged.

Indianapolis Zoo.   1200 W. Washington St. (317) 630-2030. This zoo has the state's largest aquarium, enclosed whale and dolphin pavilion, and more than 2000 animals. Admission charged.

Union Station.   39 W. Jackson Pl. (317) 267-0701. Fully restored marketplace with more than 75 stores. Free.

Other Attractions

Indiana Pacers.   Market Square Arena, 300 E. Market Sq. (317) 263-2100. Professional basketball.

Indianapolis Colts.   RCA Dome, 100 S. Capitol Ave. (317) 297-2658. Professional football.

Indianapolis Motor Speedway and Hall of Fame Museum.   4790 W. 16th St. (317) 484-6747. Museum has antique and classic passenger cars exhibits and more than 30 Indianapolis winning racecars. Admission charged.

Interesting Facts & Events

Sunday before Memorial Day. *Indianapolis 500.*

Early mid-August. *Indiana State Fair.* Fairgrounds. (317) 927-7500 or (317) 923-3431.

For More Information Contact the Convention and Visitor's Association at 1 RCA Dome, Suite 100, (317) 639-4282, for more information, a tourist package, and discounts.

Lodging

Unless otherwise noted, all lodging is within 30 minutes of the primary airport. (Rates are usually higher during Indianapolis 500 and state fair; there may be a 3-day minimum.)

Canterbury Hotel
123 S. Illinois, downtown
(317) 634-3000
Rates: $175–$225

Comfort Inn
5855 Rockville Rd., near
International Airport
(317) 487-9800
Free airport transportation.
Rates: $79–$185

Courtyard by Marriott–Airport
5525 Fortune Circle E., at
International Airport
(317) 248-0300
Free airport transportation.
Rates: $89–$120

Crown Plaza–Union Station
123 W. Louisiana St., at
Union Station
(317) 631-2221
Free airport transportation.
Rates: $125–$250

Embassy Suites
110 W. Washington St.
(317) 236-1800
Rates: $139–$219

Signature Inn
4402 E. Creek View Dr.
(317) 784-7006
Rates: $65–$72

Restaurants

Hollyhock Hill
8110 N. College Ave.
(317) 251-2294
Specializes in fried chicken.
Dinner: $13–$18

Restaurant at the Canterbury
The Canterbury Hotel
123 S. Illinois
(317) 634-3000
Specializes in seafood and veal.
Dinner: $17–$30

St. Elmo Steak House
127 S. Illinois, downtown
(317) 637-1811
Specializes in steak and
fresh seafood.
Dinner: $17–$35

Maine

Augusta

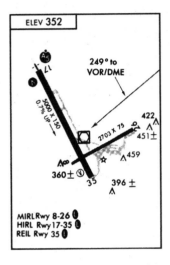

ELEV 352

249° to VOR/DME

5000 X 150
0.7% UP

2703 X 75

422
451±
459
360± ⑨
35
396 ±

MIRL Rwy 8-26 ●
HIRL Rwy 17-35 ●
REIL Rwy 35 ●

Augusta, the capital of Maine, was founded in 1628 as a trading post. Located 39 miles from the sea, Augusta is at the peak of navigation on the Kennebec River.

Airport Augusta State (AUG) is located 1 mile northwest of the city. Coordinates: N44° 19.24' W069° 47.84'.

Traffic Pattern The traffic pattern is flown at 1203 ft MSL. Lefthand pattern for Runways 8, 17, 26, and 35. Runways 8 and 26 are asphalt surfaced, 2703 ft long by 75 ft wide. Runways 17 and 35 are asphalt surfaced, 5000 ft long by 150 ft wide.

FBO Maine Instrument Flight. (207) 622-7858. Hours: 7 A.M.–dark. Frequency 123.0.

Navigational Information AUG is located on the Montreal Sectional Chart or L26 low en route chart. AU-366 VOR 171° at 6.1 miles.

Instrument Approaches ILS, VOR, and VOR/DME approaches are available.

Fuel 100LL and JetA are available at the FBO. Exxon and most credit cards are accepted.

Frequencies
CLEARANCE DELIVERY 119.95   APPROACH PORTLAND 128.35, BOSTON CENTER 128.2, BOSTON CENTER 124.25   UNICOM/CTAF 123.0.

Runway Lights Operated by the tower.

Transportation

Rental Cars
Budget, (207) 623-0210

Taxicabs
Airport, (207) 441-6235

Courtesy Cars The FBO has one available.

Area Attractions

Blaine House. State and Capitol Sts. (207) 287-2301. House of James G. Blaine, former speaker of the U.S. House of Representatives. Built in 1833, it is the official residence of Maine's governors. Free.

Maine State Museum. State and Capitol Sts. (207) 287-2301. Displays Maine's natural environment, prehistory, social history, and manufacturing heritage. Admission charged.

State House. State and Capitol Sts. (207) 287-2301. Built in 1829–1832, on its 185-foot dome is a statue of a classically robed woman bearing a pine bough torch. Free.

Other Attractions

Old Fort Western. City Center Plaza, 16 Cony St. (207) 626-2385. Built in 1754 by Boston merchants. Personnel in Colonial dress depict 18th-century life on the Kennebec River. Admission charged.

Interesting Facts & Events

Ten days in late June–early July. *Great Whatever Family Festival Week.* Events include tournaments, carnival, barbecue, parade, and fireworks.

For More Information Contact the Kennebec Valley Chamber of Commerce at University Dr., (207) 623-4559, for more information, a tourist package, and discounts.

Lodging

Unless otherwise noted, all lodging is within 30 minutes of the primary airport.

Augusta
390 Western Ave.
(207) 622-6371
Rates: $89–$169

Best Western Senator Inn
284 Western Ave.
(207) 622-5804
Rates: $79–$109

Comfort Inn
281 Civic Center Dr.
(207) 623-1000
Rates: $60–$125

Motel 6
18 Edison Dr.
(207) 622-0000
Rates: $33–$39

Bed & Breakfast

Wings Hill
Belgrade Lakes
(207) 495-2400
Rates: $95

Bangor

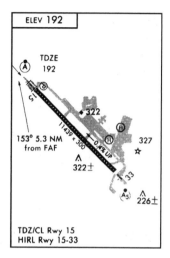

ELEV 192

TDZE
Ⓐ 192

153° 5.3 NM
from FAF

322

327

322±

A5 226±

TDZ/CL Rwy 15
HIRL Rwy 15-33

Settled in 1604, Bangor is the third largest city in Maine. Originally a harbor town, it has since evolved into a distribution center.

Airport Bangor (BGR) is located 3 miles west of the city. Coordinates: N44° 48.45' W068° 49.69'.

Traffic Pattern The traffic pattern is flown at 1200 ft MSL. Lefthand pattern for Runways 15 and 33. Runways 15 and 33 are concrete surfaced, 11,439 ft long by 300 ft wide.

FBO Bangor Aviation Services. (207) 945-5998. Hours: 24. Frequency 122.95.

FBO Telford Aviation, Inc. (207) 872-5555. Hours: 6 A.M.–9 P.M. Frequency 122.95.

Navigational Information BGR is located on the Halifax Sectional Chart or L26 low en route chart. From the BGR 114.8 VOR 155° at 2.8 miles.

Instrument Approaches ILS, VOR, VOR/DME, NDB, ASR, and GPS approaches are available.

Cautions Deer are reported in the vicinity.

Fuel 100LL and JetA are available 24 hours at Bangor Aviation Services. Exxon credit cards are accepted.

Frequencies
TOWER 120.7 (24 HR.)   GROUND 121.9   CLEARANCE DELIVERY 135.9   APPROACH BANGOR 124.5   UNICOM 122.95   ATIS 127.75.

Runway Lights Operated by the tower.

Transportation

Rental Cars
Avis, (207) 947-8383
Budget, (207) 945-9429
Hertz, (207) 942-5519
National, (207) 947-0158

Taxicabs
Airport, (207) 947-8294
Airport Shuttle, (207) 223-4070
Dick's Taxi, (207) 942-6403
Town Taxi, (207) 945-5671

Courtesy Cars   None available.

Area Attractions

Bangor Historical Museum.   159 Union at High St. (207) 942-5766. Gallery features changing exhibits. Admission charged.

Cole Land Transportation Museum.   405 Perry Rd. (207) 990-3600. More than 200 vehicles are exhibited. Admission charged.

Monument to Paul Bunyan.   Main St. in Bass Park. A 31-ft tall lumberjack statue. Free.

Interesting Facts & Events

Mid-April.   *Kenduskeag Stream Canoe Race.* (207) 947-1018.

Late June–First week of August.   *Bangor Fair.* Fair events, horse racing, and exhibits. (207) 942-9000.

Harness Racing.   Bass Park. (207) 866-7650.

Band Concerts.   Paul Bunyan Park. (207) 947-1018.

For More Information   Contact the Greater Bangor Chamber of Commerce at (207) 947-0307 for more information, a tourist package, and discounts.

Lodging

Unless otherwise noted, all lodging is within 30 minutes of the primary airport.

Best Western White House
155 Littlefield Ave.
(207) 862-3737
Rates: $59–$94

Comfort Inn
750 Hogan Rd.
(207) 942-7899
Free airport transportation.
Rates: $59–$79

Days Inn
250 Odin Rd.
(207) 942-8272
Free airport transportation.
Rates: $50–$80

Holiday Inn–Civic Center
500 Main St.
(207) 947-8651
Free airport transportation.
Rates: $100–$150

Marriott–Bangor Airport
308 Godfrey Blvd.
(207) 947-6721
Airport transportation.
Rates: $98–$155

Bed & Breakfast

Lucerne
E. Holden
(207) 843-5123
Rates: $99–$129

Phenix Inn at West Market Square
20 Broad St.
(207) 947-0411
Rates: $65–$90

Restaurants

Greenhouse
193 Broad St.
(207) 945-4040
Specializes in seafood.

Miller's
427 Main St.
(207) 945-5663
Specializes in prime rib.

Pilots Grill
1528 Hammond St.
(207) 942-6325
Specializes in baked stuffed lobster.

Portland

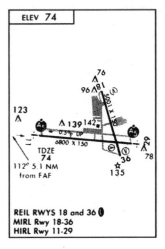

ELEV 74

REIL RWYS 18 and 36
MIRL Rwy 18-36
HIRL Rwy 11-29

Portland is Maine's largest city. Situated on Casco Bay, fishing has remained Portland's most important industry. A city of large elms, stately old homes, and historic churches, Portland abounds in historical significance dating back to the Revolution.

Airport Portland International Jetport (PWM) is located 2 miles west of the city. Coordinates: N43° 38.77' W070° 18.52'.

Traffic Pattern The traffic pattern is flown at 1074 ft MSL. Lefthand pattern for Runways 11, 18, 29, and 36. Runways 11 and 29 are asphalt surfaced, 6800 ft long by 150 ft wide. Runways 18 and 36 are asphalt surfaced, 5001 ft long by 150 ft wide.

FBO Irving Aviation Services. (207) 775-5635. Hours: 5 A.M.–11 P.M. Frequency 123.50.

FBO Northeast Air, Inc. (207) 774-6318. Hours: 24. Frequency 122.95.

Navigational Information PWM is located on the New York Sectional Chart or L26 low en route chart. From the ENE 117.1 VOR 062° 18.8 miles.

Instrument Approaches ILS, ILS/DME, and NDB approaches are available.

Cautions There are reports of sea gulls, deer, and other wildlife in the vicinity.

Fuel 100LL and JetA are available. Exxon and most major credit cards are accepted.

Frequencies
TOWER 120.9   GROUND 121.9   CLEARANCE DELIVERY 121.65
APPROACH PORTLAND 119.75, PORTLAND 125.5, BOSTON
CENTER 128.2   UNICOM 122.95   CTAF 120.9   ATIS 119.05.

Runway Lights   Operated by the tower.

Transportation

Rental Cars
Avis, Budget, Hertz, and *National* are available at the terminal.

Taxicabs
Available at the terminal.

Courtesy Cars   The terminal and Northeast Air, Inc., have one available.

Area Attractions

Bay View Cruises.   Fisherman's Wharf, 184 Commercial St. (207) 761-0496. Cruise along Casco Bay with stops at individual islands and other locations.

Children's Museum of Maine.   142 Free St. (207) 828-1234. This hands-on museum allows children to become lobster trappers, storekeepers, bankers, computer experts, and astronauts. Admission charged.

Old Port Exchange.   Exchange and Pearl Sts. Shops and restaurants are located in 19th-century brick buildings.

Portland Museum of Art.   7 Congress Sq. (207) 775-6148. American and European painting, sculpture, prints, and decorative art are displayed, including artists associated with Maine. Admission charged.

Southworth Planetarium.   96 Falmouth St. (207) 780-4249. Astronomy shows, laser light concerts, and children's shows. Admission charged.

Other Attractions

The Museum of Portland Headlight.   Cape Elizabeth. (207) 799-2661. The first lighthouse (1791) authorized by the United States and the oldest in continuous use.

Tate House. 1270 Westbrook St. (207) 774-9781. Furnished and decorated in the period 1755–1800, with 18th-century herb gardens. Admission charged.

Two Lights State Park. Cape Elizabeth. (207) 799-5871. Located on the Atlantic Ocean, an ideal fishing and picnicking spot.

Wadsworth–Longfellow House. 487 Congress St. (207) 772-1807. Boyhood home of Henry Wadsworth Longfellow. Built in 1785, it displays furnishings, portraits, and personal possessions of the family.

Interesting Facts & Events

Early June. *Old Port Festival.* Exchange St. (207) 772-6828.

Late June–August. *Outdoor Summer Concerts.* Deering Oaks Park. (207) 874-8793.

Third Saturday in August. *Sidewalk Art Show.* Congress St. to Monument Sq. (207) 828-6666.

For More Information Contact the Convention and Visitor's Bureau of Greater Portland, (207) 772-5800, for more information, a tourist package, and discounts.

Lodging

Unless otherwise noted, all lodging is within 30 minutes of the primary airport.

Best Western Merry Manor Inn
700 Main St.
(207) 774-6151
Rates: $90–$100

Holiday Inn by the Bay
88 Spring St.
(207) 775-2311
Rates: $128–$158

Embassy Suites
1050 Westbrook St.
(207) 775-2200
Rates: $159–$299

Inn by the Sea
40 Bowery Beach Rd.
(207) 799-3134
Rates: $180–$420

Marriott at Sable Oaks
200 Sable Oaks Dr.
(207) 871-8000
Rates: $129–$300

Bed & Breakfast

Inn at St. John
939 Congress St.
(207) 773-6481
Near International Airport.
Rates: $40–$125

Inn on Carleton
46 Carleton St.
(207) 775-1910
Rates: $65–$140

Pomegranate
49 Neal St.
(207) 772-1006
Rates: $135–$165

Restaurants

Back Bay Grill
65 Portland St.
(207) 772-8833
Specializes in fresh seafood and grilled dishes.

Boone's
6 Custom House Wharf
(207) 774-5725
Specializes in steak.

DiMillo's Floating Restaurant
25 Long Wharf
(207) 772-2216
Specializes in lobster.

The Roma
769 Congress St.
(207) 773-9873
Specializes in Northern Italian fare.

Walter's Café
15 Exchange St.
(207) 871-9258
Specializes in lobster with angel hair pasta.

Massachusetts

Hyannis

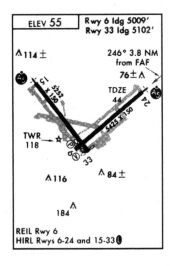

ELEV 55 | Rwy 6 ldg 5009'
Rwy 33 ldg 5102'

Λ114±

246° 3.8 NM
from FAF

76± Λ

TDZE
44

TWR
118

Λ116   Λ 84±

Λ
184

REIL Rwy 6
HIRL Rwys 6-24 and 15-33

Hyannis is the main vacation center of Cape Cod. Tennis, golf, shopping, museums, libraries—they are all here, as well as the other 6 million people who visit this village every year.

Airport Barnstable Municipal–Boardman/Polando (HYA) is located 1 mile north of the city. Coordinates: N41° 40.16' W070° 16.82'.

Traffic Pattern The traffic pattern is flown at 1052 ft MSL. Lefthand pattern for Runways 6, 15, 24, and 33. Runways 15 and 33 are asphalt surfaced, 5252 ft long by 150 ft wide. Runways 6 and 24 are asphalt surfaced, 5425 ft long by 150 ft wide.

FBO Griffin Aviation Services. (508) 771-2865. Hours: 8 A.M.–8 P.M. Frequency 122.95.

FBO Hyannis Air Service. (508) 775-8171. Hours: 6 A.M.–9 P.M. Frequency 123.3.

Navigational Information HYA is located on the New York Sectional Chart or L25 on low en route chart. From the MVY 114.5 VOR 057° 22.2 miles.

Instrument Approaches ILS, VOR, NDB, and GPS approaches are available.

Cautions Birds are reported in the vicinity.

Fuel   100LL and JetA are available. Exxon and most major credit cards are accepted.

Frequencies
TOWER 119.5   GROUND 121.9   CLEARANCE DELIVERY 125.15
APPROACH CAPE 118.2, BOSTON CENTER 132.9   UNICOM 122.95
CTAF 119.5   ATIS 123.8.

Runway Lights   Pilot-controlled lighting.

Transportation

Rental Cars   Available at the terminal.
Avis, (508) 775-2888
Budget, (508) 790-0163
Hertz, (508) 775-5825
National, (508) 771-4353

Taxicabs
Town, (508) 771-5555

Courtesy Cars   None available.

Area Attractions

Hyannis–Nantucket or Martha's Vineyard Day Round Trip.   Pier 1, Ocean St. dock. (508) 778-2600. Fishing excursions, half day or all day deep-sea fishing, or hourly sightseeing trips to Hyannis Port. Admission charged.

John F. Kennedy, Hyannis Museum.   397 Main St., in Old Town Hall. (508) 372-5230. View exhibits highlighting President Kennedy's long relationship with Cape Cod. Admission charged.

John F. Kennedy Memorial.   Ocean St. Twelve-foot high circular field-stone wall memorial to the late president.

Other Attractions

Auto Ferry Service.   South St. dock. (508) 540-2022. Nantucket Steamship Authority provides trips to Nantucket from Hyannis.

Sea St. Beach, Kalmus Park, and Veteran's Park. Swimming with bathhouse facilities along with picnicking areas. Parking fee at all beaches.

Interesting Facts & Events

Early June weekend. *Hyannis Harbor Festival.* Waterfront at Bismore Park. Annual event with boat tours, sailboat races, food, and entertainment. Call for dates. (508) 362-5230.

July–early September. *Cape Cod Melody Tent.* 21 W. Main St. Summer musical theater in-the-round. (508) 775-9100.

For More Information Contact the Chamber of Commerce at (508) 362-5230 for more information, a tourist package, and discounts.

Lodging

Unless otherwise noted, all lodging is within 30 minutes of the primary airport.

The Cape Codder Hotel
MA 132
(508) 771-3000
Rates: $99–$275

Captain Gosnold House
230 Gosnold St.
(508) 775-9111
Rates: $75–$260

Days Inn
867 Iyanough Rd.
(508) 771-6100
Rates: $110–$135

International Inn
662 Main St.
(508) 775-5600
Rates: $110–$258

Quality Inn
1470 MA 132
(508) 790-2336
Rates: $108–$118

Bed & Breakfast

Sea Breeze
397 Sea St.
(508) 771-7213
Rates: $75–$150

Simmons Homestead
288 Scudder Ave.
(508) 778-4999
Rates: $140–$260

Restaurants

Barbyann's
120 Airport Rd.
(508) 775-9795
Specializes in prime rib and seafood.

Paddock
W. Main St. at W. End Rotary
(508) 775-7677
Specializes in fresh local seafood.

Roadhouse Café
488 South St.
(508) 775-2386
Specializes in thin-crust pizza.

Sam Diego's
950 Iyanough Rd.
(508) 771-8816
Mexican menu. Specializes in fajitas and enchiladas.

Starbuck's
645 MA 132
(508) 778-6767
Specializes in hamburgers.

Nantucket

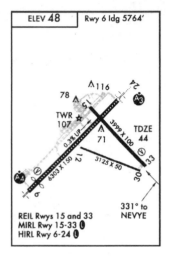

ELEV 48 | Rwy 6 ldg 5764'

REIL Rwys 15 and 33
MIRL Rwy 15-33
HIRL Rwy 6-24

Located on an island, Nantucket has 49 square miles of beaches just south of Cape Cod. Although it was the world's greatest whaling port during the 17th century, today visitors find a great variety of beaches, swimming, tennis, golf, fishing, and sailing activities.

Airport Nantucket Memorial (ACK) is located 3 miles southeast of the city. Coordinates: N41° 15.18' W070° 03.61'.

Traffic Pattern The traffic pattern is flown at 848 ft MSL. Lefthand pattern for Runways 6, 12, 15, 24, 30, and 33. Runways 12 and 30 are asphalt surfaced, 3125 ft long by 50 ft wide (daylight use only). Runways 15 and 33 are asphalt surfaced, 3999 ft long by 150 ft wide. Runways 6 and 24 are asphalt surfaced, 6303 ft wide by 150 ft wide.

FBO Graylady Aviation. (508) 228-5888. Hours: 9 A.M.–5 P.M. Also offers repair service. Frequency 122.95.

FBO Nantucket Memorial Airport. (508) 325-5307. Hours: daylight. Frequency 122.95.

Navigational Information ACK is located on the New York Sectional Chart or L25 low en route chart. From ACK 116.2 VOR 236° at 2.3 miles.

Instrument Approaches ILS, LOC, BC, VOR, NDB, and GPS approaches are available.

Cautions Deer are reported in the vicinity.

Fuel   100LL and JetA are available. Phillips, Texaco, and most major credit cards are accepted.

Frequencies
TOWER 118.3   GROUND 121.7   CLEARANCE DELIVERY 128.25
APPROACH CAPE 126.1, BOSTON CENTER 132.9   UNICOM 122.95
CTAF 118.3   ATIS 126.6.

Runway Lights   Pilot-controlled lighting.

Transportation

Rental Cars   Available at the terminal.
Budget, (508) 228-5666
Hertz, (508) 228-9421
Thrift Car, (508) 325-4616
Windmill, (508) 228-1227

Taxicabs
Available at the terminal.

Courtesy Cars   None available.

Area Attractions

Jethro Coffin House.   Sunset Hill. Nantucket's oldest house (1686). Admission charged.

Old Fire Hose Cart House.   Gardner St. off Main St. Houses firefighting equipment from bygone days. Built in 1886. Free.

Old Windmill.   Mill Hill off Prospect St. Original machinery and grinding stones housed in the windmill built in 1746 from wrecked vessels. Admission charged.

Whaling Museum.   Broad St. near Steamboat Wharf. Great collection of relics from whaling days. Admission charged.

Other Attractions

Gail's Tours.   Information Center at Federal and Broad. (508) 257-6557. Three narrated van tours daily. Admission charged.

Interesting Facts & Events

Last weekend in April. *Daffodil Festival.* Parade of antique cars. (508) 228-1700.

Second weekend in June. *Harborfest.*

Third Sunday in August. *Sand Castle Contest.* (508) 228-1700.

For More Information Contact the Chamber of Commerce at (508) 228-1700 for more information, a tourist package, and discounts.

Lodging

Unless otherwise noted, all lodging is within 30 minutes of the primary airport.

Harbor House
S. Beach St.
(508) 228-1500
Rates: $235–$285

Nantucket Inn
27 Macy's Ln.
(508) 228-6900
Rates: $130–$190

Wharf Cottages
Commercial St., foot of Main St.
(508) 228-4620
Rates: $264–$578

White Elephant Inn and Cottages
Easton St.
(508) 228-2500
Rates: $264–$695

Bed & Breakfast

Carlisle House
26 N. Water St.
(508) 228-0720
Rates: $65–$175

Carriage House
5 Ray's Ct.
(508) 228-0326
Rates: $120–$150

Jared Coffin House
29 Broad St.
(508) 228-2400
Rates: $74–$200

Sherburne Inn
10 Gay St.
(508) 228-4425
Rates: $125–$175

The Wauwinet
Wauwinet Rd.
(508) 228-0145
Rates: $320–$710

Restaurants

Cap'n Tobey's Chowder House
Straight Wharf
(508) 228-0836
Specializes in clam chowder and
scallops.

Tavern at Harbor Square
Straight Wharf off Main St.
(508) 228-1266
Specializes in New England
clam chowder.

Le Languedoc
24 Broad St.
(508) 228-2552
Specializes in veal.

Topper's
At The Wauwinet
(508) 228-0145
Specializes in seafood.

Rope Walk
Straight Wharf
(508) 228-8886
Specializes in fresh local seafood.

21 Federal
21 Federal St.
(508) 228-2121
Specializes in seafood.

Plymouth

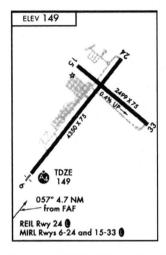

ELEV 149

TDZE 149

057° 4.7 NM
from FAF

REIL Rwy 24
MIRL Rwys 6-24 and 15-33

Plymouth was settled in 1620 by families arriving on the *Mayflower*, founding the first permanent European settlement in America north of Virginia. Plymouth Rock now lies under a granite colonnade, marking the place of landing.

Airport Plymouth Municipal (PYM) is located 4 miles southwest of the city. Coordinates: N41° 54.59' W070° 43.73'.

Traffic Pattern The traffic pattern is flown at 1150 ft MSL. Lefthand pattern for Runways 6, 15, 24, and 33. Righthand pattern for gliders and tow planes for grass next to Runways 6 and 33. Runways 15 and 33 are asphalt surfaced, 2501 ft long by 75 ft wide. Runways 6 and 24 are asphalt surfaced, 3501 ft long by 75 ft wide.

FBO Plymouth Airport. (508) 746-2020. Hours: 8 A.M.–9 P.M. daily. Frequency 123.0.

FBO Yankee Aviation Services. (508) 746-5511. Hours: 8 A.M.–5 P.M. weekdays; 8 A.M.–4:30 P.M. Saturday. Frequency 122.95.

Navigational Information PYM is located on the New York Sectional Chart or L25, L28 low en route chart. From the LFV 114.7 VOR 274° at 31.6 miles.

Instrument Approaches An NDB approach is available.

Fuel 100LL and JetA are available. Texaco and most major credit cards are accepted.

Frequencies
CLEARANCE DELIVERY 127.75   APPROACH CAPE 126.3, BOSTON CENTER 132.9   UNICOM CTAF 123.0   ASOS 135.625.

Runway Lights Pilot-controlled lighting.

Transportation

Rental Cars
Shire Town Ford, (508) 746-3400
Thrifty, (508) 747-2120
Verc, (508) 747-1997

Taxicabs
Blue Cab, (508) 746-2525
Central Cab, (508) 746-0018
Plymouth, (508) 746-0189

Courtesy Cars None available.

Area Attractions

Cape Cod Cruises. State Pier. (508) 747-2400. Historic tours of Plymouth harbor. Daily ferry service. Admission charged.

Harlow Old Fort House. Visit a typical Pilgrim household with spinning and candle-dipping demonstrations. Built in 1677. Admission charged.

Mayflower Society House Museum. 4 Winslow St. (508) 746-2590. Built in 1754, it houses the national headquarters of the General Society of Mayflower Descendants. Authentic 17th-and 18th-century furnishings. Admission charged.

Mayflower II. State Pier on Water St. A 90-ft full-size bark, reproduction of the type of ship used by the Pilgrims. Built in England and sailed to America, the *Mayflower II* is staffed by an authentically costumed crew. Admission charged.

Supersports Family Fun Park. 108 N. Main St. (508) 866-9655. Fun for the entire family: rides, games, sports, and minigolf. Admission charged.

Other Attractions

Plymouth Colony Winery. U.S. 44 W. (508) 747-3334. Tour grape vineyards and cranberry bogs. See cranberry harvest activities in the fall. Free.

Village Landing Marketplace. 170 Water St. Replica of Colonial marketplace with restaurant and shops.

Interesting Facts & Events

Each Friday in August. *Pilgrim's Progress.* From Cole's Hill to Burial Hill. Reenactment of Pilgrims going to church.

Thanksgiving Week. Various events commemorating Pilgrim Thanksgiving events. (508) 747-7525.

For More Information Contact Destination Plymouth at (800) USA-1620 for more information, a tourist package, and discounts.

Lodging

Unless otherwise noted, all lodging is within 30 minutes of the primary airport.

Blue Spruce
710 State Rd.
(508) 224-3990
Rates: $58–$70

Cold Spring
188 Court St.
(508) 746-2222
Rates: $58–$78

John Carver Inn
25 Summer St.
(508) 746-7100
Rates: $59–$115

Pilgrim Sands
150 Warren Ave.
(508) 747-0900
Rates: $88–$128

Sheraton Inn Plymouth at
Village Landing
180 Water St.
(508) 747-4900
Rates: $95–$125

Sleepy Pilgrim
182 Court St.
(508) 746-1962
Rates: $58–$68

Restaurants

Hearth and Kettle
John Carver Inn
25 Summer St.
(508) 746-7100
Specializes in fresh seafood.

McGrath's
Water St.
(508) 746-9751
Specializes in seafood and
prime rib.

Worcester

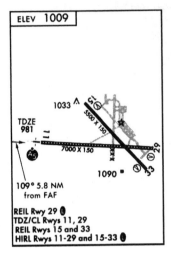

ELEV 1009

1033

TDZE 981

7000 X 150

5500 X 150

1090

109° 5.8 NM
from FAF

REIL Rwy 29
TDZ/CL Rwys 11, 29
REIL Rwys 15 and 33
HIRL Rwys 11-29 and 15-33

Settled in 1673, Worcester is known as the Heart of the Commonwealth. It is one of the largest cities in New England as well as an important industrial and cultural center.

Airport Worcester Municipal (ORH) is located 3 miles west of the city. Coordinates: N42° 16.04' W071° 52.54'.

Traffic Pattern The traffic pattern is flown at 1809 ft MSL. Lefthand pattern for Runways 11, 15, 29, and 33. Runways 11 and 29 are asphalt surfaced, 6999 ft long by 150 ft wide. Runways 15 and 33 are asphalt surfaced, 5500 ft long by 150 ft wide.

FBO DynAir Fueling, Inc. (508) 755-5870. Hours: 6 A.M.–10 P.M. Frequency 122.95.

Navigational Information ORH is located on the New York Sectional Chart of L25, L28 low en route chart. From the GDM 110.6 VOR 168° at 18.6 miles.

Instrument Approaches ILS, NDB, and GPS approaches are available.

Cautions Intensive flight training is reported in the area.

Fuel 100LL and JetA are available. Most major credit cards are accepted.

Frequencies
TOWER 120.5   GROUND 121.9   CLEARANCE DELIVERY 124.35   APPROACH BRADLEY 119.0   UNICOM 122.95   CTAF 120.5   ATIS 126.55.

Runway Lights Pilot-controlled lighting.

Transportation

Rental Cars   Available at the terminal.
Avis, (508) 754-7004 or (800) 331-1212
Hertz, (508) 753-7203 or (800) 654-3131
National, (508) 791-1345 or (800) 227-7368

Taxicabs
Arrow, (508) 756-5184
Red Cab, (508) 792-9999
Yellow, (508) 754-3211

Courtesy Cars   None available.

Area Attractions

American Antiquarian Society.   185 Salisbury St. (508) 755-5221.
Displayed is a large collection of antiques from the first 250 years of
American history. Free.

New England Science Center.   222 Harrington Way. (508) 791-
9211. Environmental science exhibits with solar/lunar observatory and
planetarium theater. Admission charged.

Salisbury Mansion.   40 Highland St. (508) 753-8278. Guided tours of
the home of Stephen Salisbury, businessman and philanthropist. Built in
1772. Admission charged.

Worcester Art Museum.   55 Salisbury St. (508) 799-4406.
Exhibitions of 50 centuries of paintings, sculpture, arts, prints, and pho-
tography from America to Egypt. Admission charged.

Other Attractions

Worcester Common Fashion Outlets.   100 Front St. (800) 2-SAVE-A-
LOT. This indoor mall contains over 100 outlet stores including a food court.

Interesting Facts & Events

September–March.   *Worcester Music Festival.* The country's oldest
music festival. (508) 754-3231.

For More Information   Contact the Worcester County Convention and Visitor's Bureau at (508) 753-2920 for more information, a tourist package, and discounts.

Lodging

Unless otherwise noted, all lodging is within 30 minutes of the primary airport.

Beechwood Hotel
363 Plantation St.
(508) 754-5789
Rates: $89–$139

Hampton Inn
110 Summer St.
(508) 757-0400
Rates: $55–$125

Days Inn
426 Southbridge
(508) 832-8300
Rates: $64–$114

Restaurants

Castle
1230 Main St.
(508) 892-9090
Specializes in breakfast.

Michigan

Detroit

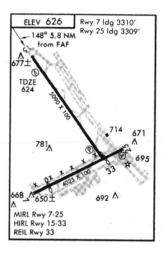

ELEV 626

Rwy 7 ldg 3310'
Rwy 25 ldg 3309'

148° 5.8 NM
from FAF

677±

TDZE 624

5090 x 100

714   671

781

695

33

4025 x 100

668   650±   692

MIRL Rwy 7-25
HIRL Rwy 15-33
REIL Rwy 33

Detroit is the city that put America on wheels. An industrial giant, it is home of the automobile industry.

Airport Detroit City (DET) is located 5 miles northeast of the city. Coordinates: N42° 24.55' W083° 00.59'.

Traffic Pattern The traffic pattern is flown at 1526 ft MSL. Lefthand pattern for Runways 7, 15, 25, and 33. Runways 15 and 33 are asphalt surfaced, 5090 ft long by 100 ft wide. Runways 7 and 25 are asphalt surfaced, 4025 ft long by 100 ft wide.

FBO Great Lakes Pilot Shop. (313) 527-7090. Hours: 8 A.M.–7 P.M. daily. Frequency 122.95.

FBO Signature Flight Support. (313) 527-6620. Hours: 24. Frequency 122.95.

FBO Wind Spirit Air, Inc. (313) 571-0060. Hours: 24. Frequency 122.95.

Navigational Information DET is located on the Detroit Sectional Chart or L12, L23 low en route chart. From the YPG 113.8 VOR 326° at 12.5 miles.

Instrument Approaches ILS, VOR, and NDB approaches are available.

Cautions Unlighted tower 275 ft AGL 1 mile S.W.

Fuel 100LL and Jet A are available. Air BP and most major credit cards are accepted.

Frequencies
TOWER 121.3   GROUND 121.9   APPROACH DETROIT 126.85
UNICOM 122.95   ATIS 133.0.

Runway Lights Operated by the tower from dusk to dawn.

Transportation

Rental Cars
Avis, (313) 371-4422

Taxicabs
Carey Limousine Service, (313) 885-5466 or (313) 923-1010
Checker, (313) 963-7000
City Cab, (313) 833-7060
Yellow Cab, (313) 961-3333

Courtesy Cars None available.

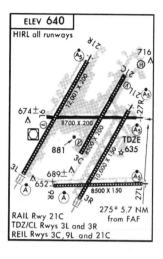

ELEV 640
HIRL all runways
716
674±
8700 X 200
TDZE
881
635
689±
652
8500 X 150
275° 5.7 NM
from FAF
RAIL Rwy 21C
TDZ/CL Rwys 3L and 3R
REIL Rwys 3C, 9L and 21C

Airport Detroit Metropolitan Wayne County (DTW) is located 15 miles south of the city. Coordinates: N42° 12.72' W083° 20.93'.

Traffic Pattern The traffic pattern is flown at 1500 ft MSL. Lefthand pattern for Runways 3L, 3C, 3R, 9, 9R, 21L, 21C, 21R, 27L, and 27. Runways 9R and 27L are concrete surfaced, 8500 ft long by 150 ft wide. Runways 3C and 21C are asphalt surfaced, 8500 ft long by 200 ft wide. Runways 3R and 21L are concrete surfaced, 10,000 ft long by 150 ft wide. Runways 3L and 21R are concrete surfaced, 12,001 ft long by 200 ft wide. Runways 9L and 27R are asphalt and concrete surfaced, 8700 ft long by 200 ft wide.

FBO Signature Flight Support. (313) 941-7880. Hours: 24. Frequency 122.95.

Navigational Information DTW is located on the Detroit Sectional Chart or L23 low en route chart. The DXO 113.4 VOR is on the field.

Instrument Approaches ILS, VOR, NDB, ASR, and GPS approaches are available.

Cautions Birds are reported in the vicinity.

Fuel 100LL and JetA are available. Air BP and most major credit cards are accepted.

Frequencies
TOWER 118.4, 135.0   GROUND 119.45, 121.8   CLEARANCE DELIVERY 120.65   DEPARTURE 118.95   APPROACH DETROIT 124.05, DETROIT 118.575, DETROIT 125.15   UNICOM 122.95 ATIS 124.55.

Runway Lights Operated by the tower.

Transportation

Rental Cars

Alamo, (800) 327-0700

Avis, (313) 942-3450 or (800) 331-1212

Budget, (313) 941-8804 or (800) 527-0700

Discount, (800) 231-7368

Dollar, (313) 942-4777 or (800) 421-6878

Enterprise, (800) 325-8007

Hertz, (313) 941-4747 or (800) 654-3131

National, (313) 941-7000 or (800) 227-7368

Thrifty, (800) 367-2277

Taxicabs

Limousine, (800) 336-4646

Metro Cars, (313) 946-5700

Taxicab, (313) 942-4690

Courtesy Cars Signature Flight Support has one available.

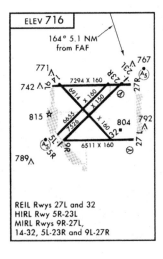

ELEV 716

164° 5.1 NM
from FAF

REIL Rwys 27L and 32
HIRL Rwy 5R-23L
MIRL Rwys 9R-27L,
14-32, 5L-23R and 9L-27R

Airport Willow Run (YIP) is located 3 miles east of the city. Coordinates: N42° 14.28' W083° 31.82'.

Traffic Pattern The traffic pattern is flown at 1716 ft MSL. Lefthand pattern for Runways 5L, 5R, 9L, 9R, 14, 23L, 23R, 27L, 27R, and 32. Runways 5L and 23R are asphalt surfaced, 6655 ft long by 160 ft wide. Runways 9R and 27L are asphalt surfaced, 6511 ft long by 160 ft wide. Runways 9L and 27R are asphalt surfaced, 7294 ft long by 160 ft wide. Runways 5R and 23L are asphalt surfaced, 7526 ft long by 150 ft wide. Runways 14 and 32 are asphalt surfaced, 6914 ft long by 160 ft wide.

FBO Chrysler Pentastar Aviation. (313) 483-3531. Hours: 24. Frequency 122.95.

FBO Signature Flight Support. (313) 482-2621. Hours: 24. Frequency 122.95.

Navigational Information YIP is located on the Detroit Sectional Chart or L23 low en route chart. From the DXO 113.4 VOR 228° at 7.5 miles.

Instrument Approaches ILS, VOR, NDB, ASR, and GPS approaches are available.

Cautions Birds are reported in the vicinity.

Fuel 100LL and JetA are available. Avfuel, Air BP, and most major credit cards are accepted.

Frequencies
TOWER 120.0   GROUND 121.9   APPROACH DETROIT 118.95
UNICOM 122.95   ATIS 127.45.

Runway Lights Operated by the tower.

Transportation

Rental Cars
Avis (through Signature Flight Support), (313) 482-2621
Pentastar Aviation, (313) 483-3531

Taxicabs
Available at the terminal.

Courtesy Cars Both FBOs have one available.

Area Attractions

Children's Museum. 67 E. Kirby Ave. (313) 494-1210. See exhibits of children's art, folk crafts, birds and mammals of Michigan, and more. Free.

Detroit Historical Museum. 5401 Woodward Ave. (313) 833-1805. History exhibits, costumes, a new automotive exhibition. Admission charged.

Detroit Zoo.   10 miles N. via Woodward Ave. in Royal Oak. (313) 398-0903. Exhibits with more than 1200 animals. Admission charged.

Other Attractions

Detroit Lions.   Pontiac Silverdome, 1200 Featherstone Rd. (248) 355-4131. Professional football.

Detroit Pistons.   The Palace of Auburn Hills, 3777 Lapeer Rd. (248) 377-0100. Professional basketball.

Detroit Tigers.   Tiger Stadium, Michigan Ave. at Trumbull. (313) 962-4000. Major League baseball.

Interesting Facts & Events

Early June.   *Detroit Grand Prix.* Belle Isle. Indy car race. (313) 259-7749.

August 18–31.   *Michigan State Fair.* Michigan Exposition and Fairgrounds. (313) 369-8250.

December.   *Christmas Carnival.* Cobo Conference/Exhibition Center. (313) 887-8200.

For More Information Contact the Metropolitan Detroit Convention and Visitor's Bureau at 100 Renaissance Center, (800) DETROIT, for more information, a tourist package, and discounts.

Lodging

Unless otherwise noted, all lodging is within 30 minutes of the primary airport.

Courtyard by Marriott
17200 N. Laurel Park Dr.
(313) 462-5907
Rates: $99–$135

Hampton Inn
30847 Flynn Dr.
(313) 721-1100
Free airport transportation.
Rates: $70–$90

Hilton Suites
8600 Wickham Rd.
(313) 728-9200
Free airport transportation.
Rates: $130–$140

Holiday Inn
5801 Southfield Service Dr.
(313) 336-3340
Rates: $99–$250

Quality Inn
7600 Merriman Rd.
(313) 728-2430
Free airport transportation.
Rates: $55–$90

The River Place Hotel
1000 River Pl. in Rivertown
(313) 259-9500 or
(800) 890-9505
Rates: $115–$500

Restaurants

Baron's Steakhouse
The River Place Hotel
1000 River Pl.
(313) 259-4855
Specializes in beef, chicken, and seafood.
Dinner: $14–$29

Joe Muer's
2000 Gratiot Ave.
(313) 567-1088
Specializes in seafood.
Dinner: $20–$26

Rattlesnake Club
300 Stroh River Pl.
(313) 567-4400
Specializes in seasonal dishes.
Dinner: $15–$30

Grand Rapids

ELEV 794

Airport Kent County International (GRR) is located 6 miles southeast of the city. Coordinates: N42° 52.96' W085° 31.44'.

Traffic Pattern The traffic pattern is flown at 1590 ft MSL. Lefthand pattern for Runways 8L, 8R, 26L, and 26R. Runways 8L and 26R are asphalt surfaced, 5000 ft long by 100 ft wide. Runways 8R and 26L are asphalt surfaced, 10,000 ft long by 150 ft wide.

FBO AMR COMBS Grand Rapids. (616) 336-4700. Hours: 6 A.M.–11 P.M. daily, after on request. Frequency 122.95.

FBO Rapid Air, Inc. (616) 957-5050. Hours: 8 A.M.–9 P.M. Frequency 122.95.

Navigational Information GRR is located on the Chicago Sectional Chart or L23, L12 low en route chart. From the GRR 110.2 VOR 349° at 5.9 miles.

Instrument Approaches ILS, VOR, NDB, ASR, and GPS approaches are available.

Fuel 100LL and JetA are available. Phillips, Air BP, and most major credit cards are accepted.

Frequencies
TOWER 135.65   GROUND 121.8   CLEARANCE DELIVERY 119.3
DEPARTURE 124.6   APPROACH GRAND RAPIDS 124.6, GRAND
RAPIDS 128.4, CHICAGO CENTER 134.95   UNICOM 122.95   CTAF
135.65   ATIS 127.1.

Runway Lights Operated by the tower. Pilot-controlled lighting midnight–5:30 A.M.

Transportation

Rental Cars
Avis, (616) 949-1720
Budget, (616) 942-1905
Hertz, (616) 949-4410
National, (616) 949-3510

Taxicabs
Calder City, (616) 454-8080

Courtesy Cars None available.

Area Attractions

Frederik Meijer Gardens. 3411 Bradford N.E. (616) 957-1580. Botanical garden and sculpture park including glass conservatory, desert garden, indoor and outdoor gardens, and more. Admission charged.

John Ball Zoo. 1300 W. Fulton St. (616) 336-4300. Children's zoo, aquarium and conservatory, and animal exhibits. Admission charged.

Other Attractions

Grand Rapids Art Museum. 155 Division St. N. (616) 456-4677. Renaissance, German, Expressionist, French, and American collections. Admission charged.

Interesting Facts & Events

First full weekend in June. *Festival.* Calder Plaza. Arts and crafts shows.

For More Information Contact the Grand Rapids/Kent County Convention and Visitor's Bureau at 140 Monroe Center N.W. (616) 459-8287 or (800) 678-9859, for more information, a tourist package, and discounts.

Lodging

Unless otherwise noted, all lodging is within 30 minutes of the primary airport.

Amway Grand Plaza
Pearl St. at Monroe Ave.
(616) 774-2000
Airport Transportation.
Rates: $91–$1100

Crowne Plaza
5700 28th St. S.E.
(616) 957-1770
Free airport transportation.
Rates: $90–$255

Best Western Midway
4101 28th St. S.E.
(616) 942-2550
Free airport transportation.
Rates: $82–$102

Holiday Inn Airport East
3333 28th St. S.E.
(616) 949-9222
Free airport transportation.
Rates: $85

Restaurants

Gibson's
1033 Lake Dr.
(616) 774-8535
Specializes in beef, lamb, and fowl.
Dinner: $12–$25

Pietro's
2780 Birchcrest St. S.E.
(616) 452-3228
Specializes in fresh pasta and chicken.
Dinner: $7–$11

Sayfee's
3555 Lake Eastbrook Blvd. S.E.
(616) 949-5750
Specializes in steak and seafood.
Dinner: $7–$19

Kalamazoo

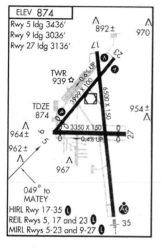

ELEV 874
Rwy 5 ldg 3436'
Rwy 9 ldg 3036'
Rwy 27 ldg 3136'

TWR 939
TDZE 874
964±
962±
967
049° to MATEY
HIRL Rwy 17-35
REIL Rwys 5, 17 and 23
MIRL Rwys 5-23 and 9-27

892±   970
954±

Airport Kalamazoo/Battle Creek International (AZO) is located 3 miles southeast of the city. Coordinates: N42° 14.09' W085° 33.12'.

Traffic Pattern The traffic pattern is flown at 1670 ft MSL. Lefthand pattern for Runways 5, 9, 17, 23, 27, and 35. Runways 17 and 35 are asphalt surfaced, 6499 ft long by 150 ft wide. Runways 5 and 23 are asphalt surfaced, 3999 ft long by 100 ft wide. Runways 9 and 27 are asphalt surfaced, 3351 ft long by 150 ft wide.

FBO Kal-Aero, Inc. (616) 343-2548. Hours: 24. Frequency 122.95.

Navigational Information AZO is located on the Chicago Sectional Chart or L23 low en route chart. The AZO 109.0 VOR is on the field.

Instrument Approaches ILS, LOC BC, VOR, NDB, ASR, and GPS approaches are available.

Cautions Birds and deer are reported in the vicinity.

Fuel 100LL and JetA are available. Avfuel and most major credit cards are accepted.

Frequencies
TOWER 118.3   GROUND 121.9   CLEARANCE DELIVERY 121.75
APPROACH KALAMAZOO 121.2, KALAMAZOO 119.2, CHICAGO CENTER 127.55   UNICOM 122.95   CTAF 118.3   ATIS 127.25.

Runway Lights Tower and pilot-controlled lighting.

Transportation

Rental Cars
Avis, (616) 381-0555
Budget, (616) 381-0617
Hertz, (616) 382-4903
National, (616) 382-2820

Taxicabs
Rapid Transit, (616) 349-9300
Yellow Cab, (616) 345-0177

Courtesy Cars   None available.

Area Attractions

Kalamazoo Air Zoo.   3101 E. Milham Rd. (616) 382-6555. Restored aircraft of World War II period, exhibits, video theater, flight simulator, and observation deck. Admission charged.

Kalamazoo Valley Museum.   230 N. Rose St. (616) 373-7990. Interactive theater, science gallery, artifacts, planetarium, and more. Admission charged.

Other Attractions

Echo Valley.   8495 East H Ave. (616) 349-3291 or (616) 345-5892. Tobogganing and ice-skating. Admission charged.

Gilmore-CCCA Museum.   6865 Hickory Rd. (616) 671-5089. More than 120 antique automobiles. Admission charged.

Interesting Facts & Events

March.   *Maple Sugaring Festival.* Kalamazoo Nature Center, 7000 N. Westnedge Ave. (616) 381-1574.

August.   *Kalamazoo County Fair.*

For More Information   Contact the Kalamazoo County Convention and Visitor's Bureau at 128 N. Kalamazoo Mall, (616) 381-4003 or (800) 222-6363, for more information, a tourist package, and discounts.

Lodging

Unless otherwise noted, all lodging is within 30 minutes of the primary airport.

Holiday Inn–Airport
3522 Sprinkle Rd.
(616) 381-7070
Free airport transportation.
Rates: $75–$83

Radisson Plaza–Kalamazoo Center
100 W. Michigan Ave.
(616) 343-3333
Free airport transportation.
Rates: $99–$275

Residence Inn by Marriott
1500 E. Kilgore Rd. near
Municipal Airport
(616) 349-0855
Free airport transportation.
Rates: $105–$130

Bed & Breakfast

Stuart Avenue
229 Stuart Ave.
(616) 342-0230
Rates: $49–$250

Restaurants

Black Swan
3501 Greenleaf Blvd.
(616) 375-2105
Specializes in fish and beef Wellington.
Dinner: $15–$20

Webster's
Radisson Plaza–Kalamazoo Center Hotel
100 W. Michigan Ave.
Specializes in beef Wellington and seafood.
Dinner: $13–$22

Saginaw

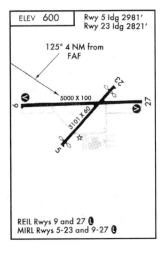

ELEV 600 | Rwy 5 ldg 2981'
Rwy 23 ldg 2821'

125° 4 NM from FAF

5000 X 100

3101 X 60

REIL Rwys 9 and 27
MIRL Rwys 5-23 and 9-27

Saginaw is known for its industry and agriculture. It is the home of numerous General Motors plants.

Airport Harry W. Browne (3SG) is located 2 miles southeast of the city. Coordinates: N43° 25.00' W83° 51.83'.

Traffic Pattern The traffic pattern is flown at 1401 ft MSL. Lefthand pattern for Runways 5, 9, 23, and 27. Runways 5 and 23 are asphalt surfaced, 3101 ft long by 60 ft wide. Runways 9 and 27 are asphalt surfaced, 3500 ft long by 75 ft wide.

FBO Harry W. Browne Airport. (517) 758-2459. Hours: daylight. Frequency 122.8.

FBO Mel's Aircraft Maintenance. (517) 754-0820. Hours 8 A.M.–5 P.M. daily. Frequency 122.95.

FBO Northwind Aviation, Inc. (517) 893-9278. Hours: 8 A.M.–5 P.M. Frequency 122.95.

FBO Reinbold Flying Service, Inc. (517) 754-7611. Hours: on request daily. Frequency 122.95.

FBO Seeley Aircraft Service. (517) 755-7271. Hours: 10 A.M.–7 P.M. Frequency 121.95.

FBO Valley Aviation. (517) 753-4951. Hours: on request daily. Frequency 122.95.

Navigational Information 3SG is located on the Detroit Sectional Chart or L12 low en route chart. From the MBS 112.9 VOR 125° at 11 miles.

Instrument Approaches VOR/DME, NDB, and GPS approaches are available.

Cautions Two towers 2–3 miles north.

Fuel 100LL and JetA are available. Most major credit cards are accepted.

Frequencies
APPROACH SAGINAW 126.45, CLEVELAND CENTER 127.7 UNICOM/CTAF 122.8.

Runway Lights Tower and pilot-controlled lighting.

Transportation

Rental Cars
Avis, (517) 695-5333—24 hr. advanced notice required.
Schaefer Bierlein Chrysler Dealer, (800) 862-8893

Taxicabs
Yellow Cab, (517) 752-3117

Courtesy Cars None available.

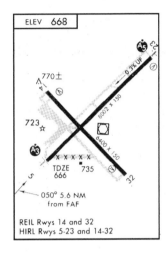

Airport MBS International (MBS) is located 9 miles northwest of the city. Coordinates: N43° 31.97' W084° 04.78'.

Traffic Pattern The traffic pattern is flown at 1660 ft MSL. Lefthand pattern for Runways 5, 14, 23, and 32. Runways 14 and 32 are asphalt surfaced, 6400 ft long by 150 ft wide. Runways 5 and 23 are asphalt surfaced, 8002 ft long by 150 ft wide.

FBO Mid Michigan AirCenter. (517) 695-2554. Hours: 5 A.M.–12 P.M. Frequency 122.95.

Navigational Information MBS is located on the Detroit Sectional Chart or L12 low en route chart. The MBS 112.9 VOR is on the field.

Instrument Approaches VOR/DME, NDB, and GPS approaches are available.

Cautions Powerlines, a tower, gulls, and deer are in the vicinity.

Fuel 100LL and JetA are available. Avfuel and most major credit cards are accepted.

Frequencies
TOWER 120.1   GROUND 121.7   CLEARANCE DELIVERY 121.85 APPROACH SAGINAW 125.46, SAGINAW 118.45, CLEVELAND CENTER 127.7   UNICOM 122.95   CTAF 120.1   ATIS 118.6.

Runway Lights Operated by the tower.

Transportation

Rental Cars
Avis, (517) 695-5333
Hertz, (517) 695-5587

National, (517) 695-5531
Thrifty, (517) 695-5308

Taxicabs
Airport, (517) 695-6513
Country Club, (517) 781-2345
D'Elegance, (800) 743-1280
Entertainment Unlimited, (517) 781-2319
Greater Bay, (517) 894-1118
Royale, (800) 817-8905

Courtesy Cars The FBO has one available.

Area Attractions

Children's Zoo. S. Washington Ave. and Ezra Rust Dr., Celebration Square. (517) 759-1657. Small animals and train and pony rides. Admission charged.

Saginaw Art Museum. 1126 N. Michigan Ave. (517) 754-2491. Exhibits of paintings, sculpture, and other fine art. Donation.

Other Attractions

Andersen Water Park and Wave Pool. Rust Ave. and Fordney St. (517) 759-1386. Wave pool, wading pool, double water slide, and other water activities. Admission charged.

Castle Museum of Saginaw County History. 500 Federal. (517) 752-2861. Replica of a French chateau with historical collections. Admission charged.

Interesting Facts & Events

Fourth weekend in July. *Greek Festival.*

First week after Labor Day. *Saginaw County Fair.*

For More Information Contact the Saginaw County Convention and Visitor's Bureau at 901 S. Washington Ave., (517) 759-1386, for more information, a tourist package, and discounts.

Lodging

Unless otherwise noted, all lodging is within 30 minutes of the primary airport.

Four Points by Sheraton
4960 Towne Centre Rd.
(517) 790-5050
Free airport transportation.
Rates: $68–$108

Hampton Inn
2222 Tittabawassee Rd.
(517) 792-7666
Rates: $56–$67

Super 8
4848 Town Centre Rd.
(517) 791-3003
Rates: $38–$54

Bed & Breakfast

Montague
1581 S. Washington Ave.
(517) 752-3939
Rates: $55–$150

Restaurants

Holly's Landing
1134 N. Niagara St.
(517) 754-4461
Specializes in steak, seafood, and prime rib.
Dinner: $10–$18

New Hampshire

Concord

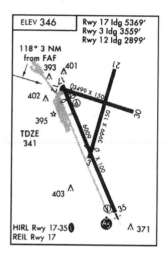

ELEV 346	Rwy 17 ldg 5369'
	Rwy 3 ldg 3559'
	Rwy 12 ldg 2899'

Concord was settled in 1727; however, New Hampshire didn't enter the Union until 1788, and Concord became its capital in 1808. With more than 400 seats, Concord has the largest legislature of any state. It remains the financial and diversified industry center of the state.

Airport Concord Municipal (CON) is located 2 miles east of the city. Coordinates: N43° 12.21' W071° 30.12'.

Traffic Pattern The traffic pattern is flown at 1346 ft MSL. Lefthand pattern for Runways 3, 12, 17, 21, 30, and 35. Runways 12 and 30 are asphalt surfaced, 3499 ft long by 150 ft wide. Runways 17 and 35 are asphalt surfaced, 6009 ft long by 100 ft wide. Runways 3 and 21 are asphalt surfaced, 3999 ft long by 150 ft wide.

FBO Concord Aviation Services. (603) 229-1760. Hours: 7 A.M.–6 P.M. weekdays, 7 A.M.–5 P.M. weekends. Frequency 122.7.

Navigational Information CON is located on the New York Sectional Chart or L25, L26 low en route chart. From CON 112.9 VOR 122° 3.4 miles.

Instrument Approaches ILS, VOR, NDB, and GPS approaches are available.

Cautions There is intensive flight training in the area.

Fuel   100LL and JetA are available. Exxon and most major credit cards are accepted.

Frequencies
CLEARANCE DELIVERY 133.65   APPROACH MANCHESTER 127.35
UNICOM/CTAF 122.7.

Runway Lights   Pilot-controlled lighting.

Transportation

Rental Cars
Avis, (603) 225-5652
Merchants, (603) 224-1300

Taxicabs
AA, (603) 225-7433
A&P, (603) 224-6573
Central, (603) 225-5569

Courtesy Cars   The FBO has one available.

Area Attractions

Canterbury Shaker Village.   I-93 Exit 18. (603) 783-9511. Museum housed in historic Shaker buildings includes crafts and inventions. Admission charged.

Christa McAuliffe Planetarium.   3 Institute Dr., I-93 Exit 15E. (603) 271-7827. In memory of the nation's first teacher in space. Admission charged.

State House.   Part St. (603) 271-2154. View the Hall of Flags, statues, and portraits of state notables. Free.

Other Attractions

Concord Arts and Crafts.   36 N. Main St. (603) 228-8171. View the displays of traditional and contemporary crafts by New Hampshire's crafts-people. Free.

Interesting Facts & Events

December–late March. Pats Peak Ski Area. NH 114 near Henniker. Enjoy winter sports. Admission charged. (603) 428-3245.

For More Information Contact the Chamber of Commerce at (603) 224-2508 for more information, a tourist package, and discounts.

Lodging

Unless otherwise noted, all lodging is within 30 minutes of the primary airport.

Brick Tower Motor Inn
414 S. Main St.
(603) 224-9565
Rates: $52–$69

Days Inn
406 S. Main St.
(603) 224-2511
Rates: $75–$95

Hampton Inn
515 South St. Bow
(603) 224-5322
Rates: $65–$120

Bed & Breakfast

Colby Hill
The Oaks, Henniker
(603) 428-3281
Rates: $75–$325

Restaurants

Colby Hill Inn
At Colby Hill Bed and Breakfast
(603) 428-3281
Specializes in chicken and seafood.

Makris Lobster and Steakhouse
Rt. 106
(603) 225-7665
Specializes in lobster.

Tio Juan's
1 Bicentennial Square
(603) 224-2821
Specializes in chimichangas, burritos, and nachos.

Manchester

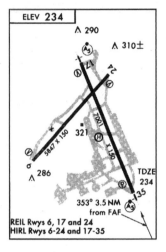

ELEV 234

△ 290

△ 310±

321

△ 286

TDZE
234

353° 3.5 NM
from FAF

REIL Rwys 6, 17 and 24
HIRL Rwys 6-24 and 17-35

Manchester is northern New England's premier financial center. When the cotton textile factory failed in 1935, the city was poverty-stricken. But the factory was bought by a group of citizens, and it revived the city.

Airport Manchester (MHT) is located 3 miles south of the city. Coordinates: N42° 56.01' W071° 26.26'.

Traffic Pattern The traffic pattern is flown at 1300 ft MSL. Lefthand pattern for Runways 6, 17, 24, and 35. Runways 17 and 35 are asphalt surfaced, 7001 ft long by 150 ft wide. Runways 6 and 24 are asphalt surfaced, 5847 ft long by 150 ft wide.

FBO Stead Aviation Corporation. (603) 669-4363. Hours: 24. Frequency 122.95.

Navigational Information MHT is located on the New York Sectional Chart or L25, L26 low en route chart. From MHT 114.4 VOR 337° at 4.9 miles.

Instrument Approaches ILS, VOR, VOR/DME, NDB, and GPS approaches are available.

Fuel 100LL and JetA are available. Exxon and most major credit cards are accepted.

Frequencies
TOWER 121.3   GROUND 121.9   CLEARANCE DELIVERY 135.9
APPROACH MANCHESTER 124.9, MANCHESTER 134.75   UNICOM
122.95   CTAF 121.3   ATIS 119.55.

Runway Lights Operated by the tower.

Transportation

Rental Cars
Avis, (603) 624-4000
Budget, (603) 668-3166
Hertz, (603) 669-6320
National, (603) 627-2999

Taxicabs
Arrow, (603) 668-1388
Best, (603) 434-6304
Executive Airport Service, (603) 625-2999

Courtesy Cars   The terminal has one available.

Area Attractions

Currier Gallery of Art.   201 Myrtle Way. (603) 669-6144. One of the leading small museums displaying 13th–20th century paintings and sculpture, New England decorative art, furniture, and glass and silver articles. Admission charged.

Science Enrichment Encounters Museum.   324 Commercial St. (603) 669-0400. Hands-on exhibits demonstrating basic science principles. Admission charged.

Other Attractions

Palace Theatre.   80 Hanover St. (603) 669-8021. Productions in a vintage vaudeville opera house.

Interesting Facts & Events

December–March.   McIntyre Ski Area. Kennard Rd. (603) 624-6571.

For More Information   Contact the Chamber of Commerce at (603) 666-6600 for more information, a tourist package, and discounts.

Lodging

Unless otherwise noted, all lodging is within 30 minutes of the primary airport.

Comfort Inn and Conference
Center
298 Queen City Ave.
(603) 668-2600
Rates: $63–$90

Econo Lodge
75 W. Hancock St.
(603) 624-0111
Rates: $40–$45

Holiday Inn–The Center
700 Elm St.
(603) 625-1000
Rates: $88–$495

Suisse Chalet Inn
860 S. Porter St.
(603) 625-2020
Rates: $56–$70

Tara Wayfarer Inn
121 S. River Rd.
(603) 622-3766
Rates: $79–$125

Bed & Breakfast

Bedford Village
2 Village Inn Ln.
(603) 472-2001
Rates: $135–$275

Restaurants

Bedford Village Inn
At Bedford Village Bed and
Breakfast
(603) 472-2001
Specializes in New England dishes.

Puritan Back Room
245 Hooksett Rd.
(603) 669-6890
Specializes in chicken tenders and
barbecued lamb.

The Yard Restaurant
1211 S. Mammoth Rd.
(603) 623-3545
Family-style dining.

Portsmouth

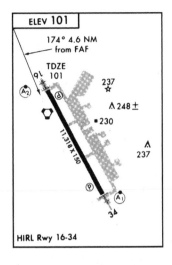

ELEV 101

174° 4.6 NM
from FAF

TDZE
101

237

A 248 ±

230

11,318 X 150

237

34

HIRL Rwy 16-34

Settled in 1630, Portsmouth has many old historic homes reflecting Colonial and Federal architecture. As the home port of many merchant sailors, a seagoing atmosphere still exists in the streets around Market Square.

Airport Pease International Trade-port (PSM) is located 1 mile west of the city. Coordinates: N43° 04.68' W070° 49.40'.

Traffic Pattern The traffic pattern is flown at 1000 ft MSL. Righthand pattern for Runway 16. Lefthand pattern for Runway 34. Runways 16 and 34 are asphalt surfaced, 11,318 ft long by 150 ft wide.

FBO SeaCoast Aviation. (603) 427-0350. Hours: 5 A.M.–11 P.M. weekdays, 6 A.M.–10 P.M. weekends. Frequency 122.95.

Navigational Information PSM is located on the New York Sectional Chart or L25, L26 low en route chart. The PSM 116.5 VOR is on the field.

Instrument Approaches ILS, ILS/DME, VOR, and GPS approaches are available.

Cautions Deer and birds are reported in the vicinity.

Fuel 100LL and JetA are available. Most major credit cards are accepted.

Frequencies
TOWER 128.4   GROUND 120.95   CLEARANCE DELIVERY 135.8
APPROACH   MANCHESTER   120.4,   BOSTON   CENTER   128.2
UNICOM 122.95   ATIS 132.05.

Runway Lights   Operated by the tower.

Transportation

Rental Cars
Ford Motor Co., (603) 427-0350
National, (603) 334-6000 or (603) 431-4707

Taxicabs
Available at the terminal.

Courtesy Cars   None available.

Area Attractions

InSight Tours.   (603) 436-4223. Specialized tours of historic Portsmouth. Admission charged.

John Paul Jones House.   43 Middle St. (603) 436-8420. Museum of the famous seafarer containing period furniture, costumes, china, glass, documents, and weapons. Built in 1758. Admission charged.

Rundlet–May House.   364 Middle St. (603) 436-3205. Built in 1807, this three-story Federal-style mansion contains terraces and its original 1812 courtyard layout. See family furnishings and Federal period crafts-manship. Admission charged.

Other Attractions

Star Island and Isles of Shoals.   Barker's Wharf, 315 Market St. (603) 431-5500. Cruise to the historic Isles of Shoals and Star Island. Whale watch expeditions and lobster clambake river cruises. Admission charged.

Interesting Facts & Events

Last Sunday in June.   Portsmouth Jazz Festival. Historical Portsmouth waterfront. (603) 436-7678.

For More Information   Contact the Greater Portsmouth Chamber of Commerce at (603) 436-1118 for more information, a tourist package, and discounts.

Lodging

Unless otherwise noted, all lodging is within 30 minutes of the primary airport.

Comfort Inn
1390 Lafayette Rd.
(603) 433-3338
Rates: $89–$115

Holiday Inn
300 Woodbury Ave.
(603) 431-8000
Rates: $106–$190

Pine Haven
183 Lafayette Rd.
(603) 964-8187
Rates: $59–$75

Suisse Chalet
650 Borthwick Ave.
(603) 436-7373
Rates: $70–$105

Bed & Breakfast

Inn at Christian Shore
335 Maplewood Ave.
(603) 431-6770
Rates: $35–$105

Restaurants

Dinner Horn Restaurant
980 Lafayette Rd.
(603) 436-0717
Specializes in seafood.

Margarita's Mexican Restaurant
755 Lafayette Rd.
(603) 431-5828
Mexican fare.

Yoken's Thar She Blows
1390 Lafayette Rd.
(603) 436-8224
Specializes in steak and local seafood.

New Jersey

Atlantic City

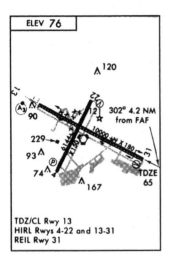

ELEV 76

A 120

A₃
90

302° 4.2 NM
from FAF

229→
93 A

74 A
P

A
167

TDZE
65

TDZ/CL Rwy 13
HIRL Rwys 4-22 and 13-31
REIL Rwy 31

Airport Atlantic City International (ACY) is located 9 miles northwest of the city. Coordinates: N39° 27.45' W074° 34.63'.

Traffic Pattern The traffic pattern is flown at 1076 ft MSL. Lefthand pattern for Runways 4, 13, 22, and 31. Runways 13 and 31 are asphalt surfaced, 10,000 ft long by 180 ft wide. Runways 4 and 22 are concrete surfaced, 6144 ft long by 150 ft wide.

FBO Midlantic Jet Aviation. (609) 383-3993. Hours: 24. Frequency 122.95.

FBO Signature Flight Support. (609) 646-5340. Hours: 24. Frequency 122.95.

Navigational Information ACY is located on the Washington Sectional Chart or L24, L28 low en route chart. The ACY 108.6 VOR is on the field.

Instrument Approaches ILS, VOR, VOR/DME, ASR, and GPS approaches are available.

Cautions Birds and deer are in the vicinity.

Fuel 100LL and JetA are available. Avfuel and most major credit cards are accepted.

Frequencies
TOWER 120.3   GROUND 121.9   CLEARANCE DELIVERY 127.85

APPROACH ATLANTIC CITY 124.6, ATLANTIC CITY 134.25 UNICOM 122.95 ATIS 108.6.

Runway Lights Operated by the tower.

Transportation

Rental Cars
Avis, (609) 345-3246
Budget, (609) 645-3494
Hertz, (609) 646-1212

Taxicabs
On request.

Courtesy Cars None available.

Atlantic City

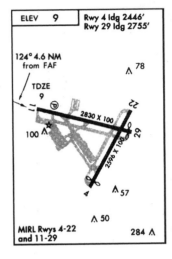

ELEV 9 | Rwy 4 ldg 2446'
Rwy 29 ldg 2755'

124° 4.6 NM
from FAF

TDZE 9

2830 X 100

2596 X 100

MIRL Rwys 4-22 and 11-29

Airport Atlantic City Municipal–Bader Field (AIY) is located 1 mile west of the city. Coordinates: N39° 21.60' W074° 27.37'.

Traffic Pattern The traffic pattern is flown at 1000 ft MSL. Lefthand pattern for Runways 4 and 11. Righthand pattern for Runways 22 and 29. Runways 11 and 29 are asphalt surfaced, 2830 ft long by 100 ft wide. Runways 4 and 22 are asphalt surfaced, 2596 ft long by 100 ft wide.

FBO None.

Navigational Information AIY is located on the Washington Sectional Chart or L24, L28 low en route chart. From the ACY 108.6 VOR 146° at 8 miles.

Instrument Approaches VOR and GPS approaches are available.

Fuel 100LL and JetA are available. Most major credit cards are accepted.

Frequencies
CLEARANCE DELIVERY 121.7   APPROACH ATLANTIC CITY 134.5, ATLANTIC CITY 124.6   UNICOM/CTAF 123.0.

Runway Lights Operated by the tower.

Transportation

Rental Cars Available at the terminal.

Taxicabs
On standby.

Courtesy Cars None available.

Area Attractions

Casinos

The Atlantic City Hilton.   Boston and Boardwalk. (800) 257-8677.

Bally's Park Place.   Park Pl. and Boardwalk. (800) 225-5977.

Caesars Atlantic City Hotel and Casino.   2100 Pacific Ave. and Boardwalk. (800) 524-2867.

Sands Hotel and Casino.   S. Indiana Ave. and Brighton Park. (609) 441-4000.

Tropicana Casino and Resort.   Brighton Ave. and Boardwalk. (800) 843-8767.

Trump Marina Hotel and Casino.   Huron Ave. and Brigantine Blvd. (800) 777-1177.

Trump Taj Mahal Casino Resort.   1000 Boardwalk. (800) 825-8786.

Absecon Lighthouse.   Pacific and Rhode Island Aves. (609) 927-5218. Designed by Civil War General George Meade and built in 1857.

Atlantic City Boardwalk.   (609) 348-7100. Built in 1870, this is the world's first boardwalk.

Atlantic City Historical Museum and Cultural Center.   Boardwalk at New Jersey Ave. (609) 347-5839. Showcases Atlantic City's long and exciting history. Free.

Ripley's Believe It or Not! Museum.   New York Ave. and Boardwalk. (609) 347-2001. Strange and unusual exhibits from Robert Ripley's travels. Admission charged.

Steel Pier.   Virginia Ave. and Boardwalk. (609) 345-4893. Rides, entertainment, games.

Other Attractions

Anglen's Choice Sportfishing Charter Service.   (609) 272-2244. Reservation service for charter boats and sportfishing.

Sea Skate Pavilion. 2301 Boardwalk. (609) 347-2414. A 17,000-ft^2 outdoor skating rink.

Interesting Facts & Events

Early February. Sail Expo Atlantic City. New Atlantic City Convention Center. (609) 449-2000.

March 17. St. Patrick's Day Parade. Atlantic City Boardwalk. (609) 641-7015.

March 28–29. Atlantic City Spring Festival. (609) 926-1800. The world's largest indoor antiques and collectibles show.

Lodging

Unless otherwise noted, all lodging is within 30 minutes of the primary airport.

The Atlantic City Hilton
Boston and Boardwalk
(800) 257-8677
Rates: $120–$200

Caesars Atlantic City Hotel and
Casino
2100 Pacific Ave. and Boardwalk
(800) 524-2867
Rates: $150–$200

Sands Hotel and Casino
S. Indiana Ave. and Brighton Park
(609) 441-4000
Rates: $100–$180

Trump Marina Hotel and Casino
Huron Ave. and Brigantine Blvd.
(800) 777-1177
Rates: $140–$200

Trump Taj Mahal Casino Resort
1000 Boardwalk
(800) 825-8786
Rates: $160–$200

Restaurants

Atlantic City Bar and Grill
1219 Pacific at S. Carolina Ave.
(609) 348-8080
Casual dining.
Dinner: $10–$15

Dock's Oyster House
2405 Atlantic Ave.
(609) 345-0092
Specializes in seafood.
Dinner: $15–$20

Flying Cloud Café
800 New Hampshire Ave.
(609) 345-8222
Specializes in seafood.
Dinner: $12–$15

Sabatini's
2210 Pacific Ave.
(609) 345-4816
Italian specialties.
Dinner: $10–$15

Ocean City

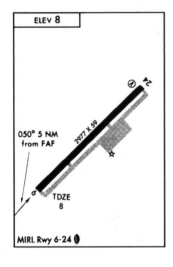

ELEV 8

050° 5 NM
from FAF

2977 X 59

TDZE
8

MIRL Rwy 6-24

People from all over the country come to this popular resort year after year. Ocean City lies between the Atlantic Ocean and Great Egg Harbor. It is known for excellent swimming, fishing, boating, golf, and tennis.

Airport Ocean City Municipal (26N) is located 2 miles southwest of the city. Coordinates: N39° 15.82' W074° 36.44'.

Traffic Pattern The traffic pattern is flown at 1007 ft MSL. Lefthand pattern for Runways 6 and 24. Runways 6 and 24 are asphalt surfaced, 2977 ft long by 59 ft wide.

FBO City of Ocean City. (609) 399-0907. Hours: 8 A.M.–4:30 P.M. winter; 8 A.M.–8 P.M. summer. Frequency 122.7.

Navigational Information 26N is located on the Washington Sectional Chart or L24, L28 low en route chart. From the SIE 114.8 VOR 051° at 13.5 miles.

Instrument Approaches VOR and GPS approaches are available.

Fuel 100LL and JetA are available. Shell and most major credit cards are accepted.

Frequencies
APPROACH ATLANTIC CITY 124.6   UNICOM/CTAF 122.7.

Runway Lights Tower and pilot-controlled lighting.

Transportation

Rental Cars
Sensible, (609) 399-0900

Taxicabs
Gerry's (609) 927-9140

Courtesy Cars   None available.

Area Attractions

Discovery Sea Shell Museum.   2721 Asbury Ave. (609) 399-1801. Over 10,000 sea shells on display. Free.

Historic House.   1139 Wesley Ave. (609) 399-1801. 1920–1930 furnishings and fashions. Open daily April–December except Sunday. Donation.

Ocean City Historical Museum.   173 Simpson Ave. at 17th St. (609) 399-1801. Featuring a doll exhibit, local shipwreck, photographic exhibit, and Victorian furnishings and fashions. Donation.

Seashore Cottage.   1139 Wesley Ave. (609) 399-1801. An example of a Victorian seaside home. Admission charged.

Other Attractions

North Star Party Boats.   9th and Palen Sts. (609) 399-7588. Marine environmental tours. Free.

Tee Time Golf.   624 Boardwalk. (609) 398-6763. Putt-putt golf. Admission charged.

Wonderland Pier.   Boardwalk and 6th St. (609) 399-7082. Mammoth Ferris wheel. Free.

Interesting Facts & Events

June.   Flower Show. Located on the Music Pier. (609) 525-9300.

August.   Boardwalk Art Show. Come see international and regional artists. (609) 525-9300.

For More Information Contact the Public Relations Department, City of Ocean City at 9th and Asbury Ave., (609) 525-9300 or (800) BEACH-NJ, for more information, a tourist package, and discounts.

Lodging

Unless otherwise noted, all lodging is within 30 minutes of the primary airport.

Beach Club Hotel
1280 Boardwalk at 13th St.
(609) 399-8555
Rates: $101–$222

Forum
8th St. and Atlantic Ave.
(609) 399-8700
Rates: $48–$134

Residence Inn by Marriott
900 Mays Landing Rd.,
Somers Point
(609) 927-6400
Rates: $160–$215

Bed & Breakfast

Serendipity
712 9th St.
(609) 399-1554
Rates: $70–$105

Terrace Bed and Breakfast
924 Ocean Blvd.
(609) 399-4246
Rates: $75–$125

Restaurants

Crab Trap
Somers Pt. Circle, Somers Point
(609) 927-7377
Family owned, specializing in stuffed lobster tail and prime rib.
Dinner: $12–$28

Deauville Inn
201 Willard Rd., Strathmore
(609) 263-2080
Outdoor dining, specializing in steak.
Dinner: $7–$23

Mac's
908 Shore Rd., Somers Point
(609) 927-4360
Specializes in seafood.
Dinner: $10–$22

Trenton

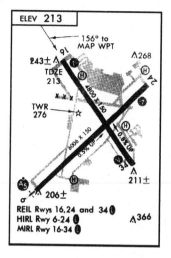

ELEV 213

156° to MAP WPT

243± TDZE 213

TWR 276

REIL Rwys 16,24 and 34
HIRL Rwy 6-24
MIRL Rwy 16-34

4800 X 150

6006 X 150

268
211±
206±
366

Trenton has been the capital of New Jersey since 1790 and is one of the fastest growing business and industrial areas in the country. Rich in history, particularly from the American Revolution, Trenton figured prominently in a Washington victory over the British. On December 25, 1776, George Washington attacked and defeated the British-held town.

Airport Mercer County (TTN) is located 4 miles northwest of the city. Coordinates: N40° 16.60' W074° 48.81'.

Traffic Pattern The traffic pattern is flown at 1200 ft MSL. Lefthand pattern for Runways 6, 16, 24, and 34. Runways 16 and 34 are asphalt surfaced, 4800 ft long by 150 ft wide. Runways 6 and 24 are asphalt surfaced, 6006 ft long by 150 ft wide.

FBO Air Hangar Inc. (609) 882-2010. Hours: 7:30 A.M.–midnight. Frequency 122.95.

FBO Ronson Aviation, Inc. (609) 771-9500. Hours: 24. Frequency 122.95.

Navigational Information TTN is located on the New York Sectional Chart or L24, L28 low en route chart. From the ARD 108.2 VOR 082° at 4.5 miles.

Instrument Approaches ILS, VOR, NDB, VOR/DME, and GPS approaches are available.

Cautions Geese and sea gulls are in the vicinity.

Fuel 100LL and JetA are available. Texaco and most major credit cards are accepted.

Frequencies
TOWER 120.7   GROUND 121.9   CLEARANCE DELIVERY 121.9
APPROACH PHILADELPHIA 123.8   UNICOM 122.95   CTAF 120.7
ATIS 133.7.

Runway Lights   Operated by the tower. Pilot-controlled lighting midnight to 6 A.M.

Transportation

Taxicabs
Available at the terminal.

Courtesy Cars   The FBOs have one available.

Area Attractions

N.J. State Archives.   185 W. State St. (609) 292-6260. Revolutionary war records and over 6000 reels of microfilmed newspapers dated as far back as the 17th century. Free.

N.J. State Museum.   205 W. State St. (609) 292-6464. Visit the planetarium on Saturday and Sunday for unique shows. The museum houses dinosaur fossils and Native American artifacts. Admission charged.

The Old Barracks Museum.   Barrack St. opposite W. Front St. (609) 396-1776. Colonial barracks built in 1758–1759 that housed British, Hessian, and Continental troops during the Revolution. Admission charged.

William Trent House.   15 Market St. (609) 989-3027. Trenton's oldest house and former home of Chief Justice William Trent, for whom the city is named. Admission charged.

Other Attractions

Trenton Thunder Baseball Team.   (609) 394-8326. Double A baseball action.

Washington Crossing State Park. 8 miles NW on NJ 29. (609) 737-0623. An 841-acre park commemorating the famous crossing on Christmas night, 1776, by the Continental Army, commanded by General George Washington. Enjoy nature trails, picnicking, and more. Also enjoy the **Ferry House State Historic Site,** (609) 737-2515, located within the park.

Interesting Facts & Events

Early May. Trenton Kennel Club Dog Show. Mercer County Central Park.

Afternoon of December 25. Reenactment of Crossing of the Delaware. Washington Crossing State Park.

For More Information Contact the Mercer County Chamber of Commerce at 214 W. State St., (609) 393-4143, for more information, a tourist package, and discounts. Web site: http://www.trentnj.com.

Lodging

Unless otherwise noted, all lodging is within 30 minutes of the primary airport.

Howard Johnson
2995 Brunswick Pike,
Lawrenceville
Rates: $63–$85

Restaurants

Diamonds
132 Kent, Chambersburg district
(609) 393-1000
Specializes in seafood.
Dinner: $13–$33

La Gondola
762 Roebling Ave.
(609) 392-0600
Italian menu.
Dinner: $13–$24

Kat-Man-Du
Rt. 29
(609) 393-7300
Specializes in American fare.
Dinner: $15–$20

New York

Buffalo

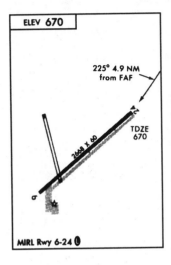

ELEV 670

225° 4.9 NM
from FAF

TDZE
670

2668 x 60

MIRL Rwy 6-24

Buffalo is New York's second largest city and one of the largest railroad centers in America.

Airport Buffalo Airfield (9G0) is located 6 miles southeast of the city. Coordinates: N42° 51.72' W078° 42.00'.

Traffic Pattern The traffic pattern is flown at 1500 ft MSL. Lefthand pattern for Runways 6 and 24. Runways 6 and 24 are asphalt surfaced, 2668 ft long by 60 ft wide.

FBO Buffalo Airfield Mgt. Corp. (716) 668-4900. Hours: daylight. Frequency 122.8.

Navigational Information 9G0 is located on the Detroit Sectional Chart or L12 low en route chart. From the BUF 116.4 VOR 226° at 5.1 miles.

Instrument Approaches VOR and GPS approaches are available.

Fuel 100LL and JetA are available. Most major credit cards are accepted.

Frequencies
APPROACH BUFFALO 123.8   UNICOM/CTAF 122.8.

Runway Lights Tower and pilot-controlled lighting.

Transportation

Rental Cars

Agency, (800) 321-1972
Snappy, (716) 683-6750
Thrifty, (800) 527-0202

Taxicabs

South Buffalo, (716) 822-6300
Radio, (716) 633-4200

Courtesy Cars   None available.

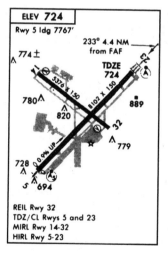

ELEV **724**
Rwy 5 ldg 7767'

233° 4.4 NM
from FAF

774±

TDZE
724

5376 X 150

780

8102 X 150

820

889

32

779

728

694

REIL Rwy 32
TDZ/CL Rwys 5 and 23
MIRL Rwy 14-32
HIRL Rwy 5-23

Airport   Greater Buffalo International (BUF) is located 5 miles east of the city. Coordinates: N42° 56.43' W078° 43.93'.

Traffic Pattern   The traffic pattern is flown at 1800 ft MSL. Lefthand pattern for Runways 5, 14, 23, and 32. Runways 14 and 32 are asphalt surfaced, 5376 ft long by 150 ft wide. Runways 5 and 23 are asphalt surfaced, 8102 ft long by 150 ft wide.

FBO   Prior Aviation Service, Inc. (716) 633-1000. Hours: 24. Frequency 122.95.

Navigational Information   BUF is located on the Detroit Sectional Chart or L12 low en route chart. From the BUF 116.4 VOR 288° at 3.9 miles.

Instrument Approaches ILS, VOR, NDB, VOR/DME, and GPS approaches are available.

Cautions Deer are reported in the vicinity.

Fuel 100LL and JetA are available. Exxon and most major credit cards are accepted.

Frequencies
TOWER 120.5   GROUND 121.9   CLEARANCE DELIVERY 124.7
APPROACH BUFFALO 123.8, BUFFALO 126.5   ATIS 135.35.

Runway Lights Operated by the tower dusk to dawn.

Transportation

Rental Cars
Alamo, (716) 631-2821
Avis, (716) 632-1808
Budget, (716) 632-4662
Hertz, (716) 632-4772
National, (716) 632-0203

Taxicabs
Independent, (716) 663-1473 or (800) 551-9369

Courtesy Cars None available.

Area Attractions

Albright–Knox Art Gallery.   1285 Elmwood Ave. (716) 882-8700. Exhibits of 18th-, 19th-, and 20th-century American and European paintings. Admission charged.

Buffalo and Erie County Botanical Gardens.   S. Park Ave. (716) 696-3555. A 150-acre park and outdoor gardens. Free.

Buffalo and Erie County Naval and Military Park.   1 Naval Park Cove. (716) 847-1773. World War II museum. Admission charged.

Buffalo Museum of Science. 1020 Humboldt Pkwy. (716) 896-5200. Exhibits on astronomy, botany, geology, zoology, and natural sciences. Admission charged.

Frank Lloyd Wright Residences. 125 Jewett Pkwy., 118 Summit, 285 Woodward, 57 Tillinghast Pl., and 76 Soldiers Pl.

Other Attractions

Buffalo Bills. Abbott Rd. and U.S. 20. (716) 648-1800. Professional football.

Buffalo Sabres. 140 Main St. (716) 855-4100. Professional hockey.
City Hall/Observation Tower. Niagara Square, 28th floor. (716) 851-5891. Panoramic view of Lake Erie. Free.

Interesting Facts & Events

August. Erie County Fair. (716) 649-3900.

For More Information Contact the Greater Buffalo Convention and Visitor's Bureau at 617 Main St., (716) 852-0511, for more information, a tourist package, and discounts.

Lodging

Unless otherwise noted, all lodging is within 30 minutes of the primary airport.

Hampton Inn
10 Flint Rd.
(716) 689-4414
Free airport transportation.
Rates: $67–$85

Holiday Inn–Downtown
620 Delaware Ave.
(716) 886-2121
Web site: http://www.
harthotels.com
Free airport transportation.
Rates: $84–$99

Hyatt Regency Buffalo
2 Fountain Plaza
(716) 856-1234
Rates: $89–$450

Microtel
1 Hospitality Centre Way
(716) 693-8100
Rates: $40–$44

Bed & Breakfast

ASA Ransom House
10529 Main St.
(716) 759-2315
e-mail: asaransom@aol.com
Each room decorated to a
distinctive theme.
Rates: $75–$150

Restaurants

The Coachman's
10350 Main St.
(716) 759-6852
Specializes in steak.
Dinner: $9–$30

Salvatore's Italian Gardens
6461 Transit Rd., near airport
(716) 683-7990
Specializes in prime rib.
Dinner: $13–$45

Long Island

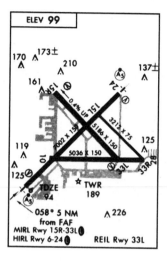

ELEV 99

170 △ 173±
△

210
△

137±
△

161
△

173±

210

137

125
△

119
△

125
△

5036 X 150

TDZE
94

☆ TWR
189

058° 5 NM
from FAF

△ 226

MIRL Rwy 15R-33L

HIRL Rwy 6-24

REIL Rwy 33L

Long Island stretches 118 miles east by northeast from the edge of Manhattan. Towns mentioned in this section include East Hampton, Fire Island National Seashore, Great Neck, Montauk, and Oyster Bay.

Airport Long Island MacArthur (ISP) is located 7 miles northeast of the city. Coordinates: N40° 47.71' W073° 06.01'.

Traffic Pattern The traffic pattern is flown at 1099 ft MSL. Lefthand pattern for Runways 6, 10, 15L, 28, and 33L. Righthand pattern for Runways 15R, 24, and 33R. Runway 15L and 33R are asphalt surfaced, 3212 ft long by 75 ft wide. Runways 15R and 33L are asphalt surfaced, 5186 ft long by 150 ft wide. Runways 10 and 28 are asphalt surfaced, 5036 ft long by 150 ft wide. Runways 6 and 24 are asphalt surfaced, 7002 ft long by 150 ft wide.

FBO Executive Fliteways, Inc. (516) 588-5454. Hours: 24. Frequency 122.95.

FBO Garrett Aviation Services. (516) 468-3000. Hours: 7 A.M.–midnight. Frequency 122.95.

FBO Long Island Jet Center. (516) 588-0303. Hours: 6 A.M.–11 P.M., 24 hours on request. Frequency 122.95.

FBO NY JET/Mid Island Air Service. (516) 588-5400. Hours: 7 A.M.–9 P.M. Frequency 123.5.

Navigational Information ISP is located on the New York Sectional Chart or L24, L25, and L28 low en route chart. From the DPK 117.7 VOR 101° at 9.3 miles.

Instrument Approaches ILS, NDB, and GPS approaches are available.

Cautions Birds are in the vicinity.

Fuel 100LL and JetA are available. Mobil, Exxon, Texaco, and most major credit cards are accepted.

Frequencies
TOWER 119.3, 124.3   GROUND 135.3   CLEARANCE DELIVERY 121.85   DEPARTURE 118.0   APPROACH NEW YORK 118.0 UNICOM 122.95   CTAF 119.3   ATIS 128.45.

Runway Lights Tower and pilot-controlled lighting.

Transportation

Rental Cars
Avis, (516) 588-6633
Budget, (800) 247-5466
Hertz, (516) 737-9200 or (800) 654-3131

Taxicabs
Airlimo, (800) 247-5466
Classic, (516) 567-5100
Colonial, (516) 589-7878
LI Airports Limo Service, (516) 234-8400
Lincoln and Limo, (516) 357-6220

Courtesy Cars All FBOs have one available.

East Hampton

Area Attractions

Guild Hall Museum.   158 Main St. (516) 324-0806. Regional art exhibits. Donation.

Hook Mill.   36 N. Main St. 1806 windmill. Admission charged.

Lodging

Unless otherwise noted, all lodging is within 30 minutes of the primary airport.

East Hampton House
226 Pantigo Rd.
(516) 324-4300
Rates: $125–$220

Bed & Breakfast

Centennial House
13 Woods Ln.
(516) 324-9414
Period furnishings.
Rates: $175–$475

Restaurants

East Hampton Point
295 Three Mile Harbor
(516) 329-2800
Specializes in seafood.
Dinner: $19–$30

Fire Island National Seashore

Area Attractions

Fire Island Lighthouse.   (516) 661-4876. Nature trail and former lightkeepers' quarters.

Sailors Haven/Sunken Forest.   (516) 589-8980. Swimming, lifeguards, marina, picnicking, and nature walks.

Smith Point West.   (516) 281-3010. Self-guided nature walk.

Watch Hill.   (516) 475-1665. Features 26 family campsites. Lottery reservations required. Phone (516) 597-6633 after May 1.

Great Neck

Area Attractions

U.S. Merchant Marine Academy.   At King's Point. (516) 773-5000. Regimental reviews and the American Merchant Marine Museum. Free.

Restaurants

Navona
218 Middle Neck Rd.
(516) 487-5603
Specializes in pasta.
Dinner: $16–$25

Montauk

Area Attractions

Montauk Point State Park. (516) 668-2554. Montauk lighthouse.

Lodging

Unless otherwise noted, all lodging is within 30 minutes of the primary airport.

Beachcomer Resort
Old Montauk Hwy.
(516) 668-2894
Rates: $155–$285

Driftwood on the Ocean
Montauk Hwy.
(516) 668-5744
Rates: $143–$295

Restaurants

Blue Marlin
240 Fort Pond Rd.
(516) 668-3249
Specializes in steak.
Dinner: $10–$19

Oyster Bay

Area Attractions

Planting Fields Arboretum. 1.5 miles W. off NY 25A. (516) 922-9201. A 400-acre estate of the late William Robertson Coe. Admission charged.

Interesting Facts & Events

Memorial Day weekend and first weekend in June.   Long Island Mozart Festival. Outdoor musical festival. (516) 922-6464.

Weekend after Columbus Day.   Oyster Festival. Street festival. (516) 922-6464.

New York

New York continues attracting more than 19 million visitors each year, and if you want to see it all, I recommend stopping at a bookstore for guidebooks that detail what to see and do.

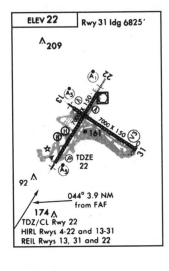

Airport La Guardia (LGA) is located 4 miles east of the city. Coordinates: N40° 46.63' W073° 52.36'.

Traffic Pattern The traffic pattern is flown at 1500 ft MSL. Lefthand pattern for Runways 4, 13, 22, and 31. Runways 4, 13, 22, and 31 are asphalt surfaced, 7000 ft long by 150 ft wide.

FBO FlightSafety International. (718) 565-4100 or (800) 227-5656. Frequency 122.95.

FBO Signature Flight Support. (718) 476-5200. Hours: 24. Frequency 122.95.

Navigational Information LGA is located on the New York Sectional Chart or L24, L25, L28 low en route chart. The LGA 113.1 VOR is on the field.

Instrument Approaches ILS, ILS/DME, LOC, VOR, VOR/DME, LDA, and NDB approaches are available.

Fuel 100LL and JetA are available. Air BP and most major credit cards are accepted.

Frequencies
TOWER 118.7 GROUND 121.7, 121.85 CLEARANCE DELIVERY 135.2 DEPARTURE 120.4 APPROACH NEW YORK 120.8, New York 132.7 NEW YORK 124.95 UNICOM 122.95 ATIS 127.05, 125.95.

Runway Lights   Operated by the tower.

Transportation

Rental Cars
Avis, (718) 507-3600
Budget, (718) 639-6400
Dollar, (718) 779-5600
Hertz, (718) 478-5300
National, (718) 672-7622

Taxicabs
Airlimo, (800) 247-5466
Carey Limousine Service, (718) 476-5353
Classic Share-Ride, (516) 567-5100
Connecticut Limo, (203) 874-6867

Courtesy Cars   Both FBOs have one available.

Area Attractions

American Museum of Natural History.   Central Park W. at 79th St. (212) 769-5650. Exhibits include the skeletal constructions of dinosaurs and other prehistoric life. Admission charged.

Carnegie Hall.   154 W. 57th St. at 7th Ave. (212) 247-7800. Celebrated auditorium. Guided 1-hour tours. Admission charged.

Central Park.   59th St. and 5th Ave. (212) 427-4040. The 843-acre park includes a lake, two skating rinks, a swimming pool, a miniature golf course, and horse-drawn carriages which can be hired for a ride through the park.

Empire State Building.   34th St. and 5th Ave. (212) 736-3100. The 1454-foot-high skyscraper has observation platforms on 86th and 102d floors. Admission charged.

Metropolitan Museum of Art.   5th Ave. at 82d St. (212) 535-7710. The most comprehensive collection in America. Admission charged.

Statue of Liberty National Monument.   Boats leave from Castle Clinton on the Battery. (212) 363-3200. Admission charged.

Other Attractions

Island Helicopter.   Departs from Heliport at 34th St. and East River. (212) 683-4575. Admission charged.

New York Giants.   Giants Stadium. (201) 935-8111. Professional football.

New York Islanders.   Nassau Veterans Memorial Coliseum. (516) 794-4100. Professional hockey.

New York Jets.   Giants Stadium. (516) 560-8100. Professional football.

New York Knickerbockers.   Madison Square Garden. (212) 465-6000. Professional basketball.

New York Mets.   Shea Stadium. (718) 507-6387. Major League Baseball.

New York Rangers.   Madison Square Garden. (212) 465-6845. Professional hockey.

New York Yankees.   Yankee Stadium. (718) 293-4300. Major League Baseball.

Interesting Facts & Events

Early mid-February.   Chinese New Year. Parade.

March 17.   St. Patrick's Day Parade. New York's biggest parade.

Weekend of July 4.   Harbor Festival.

Mid-October.   Columbus Day Parade.

Late November.   Thanksgiving Day Parade.

For More Information   Contact the New York Convention and Visitor's Bureau at 810 7th Ave., (212) 484-1200, for more information, a tourist package, and discounts.

Lodging

Unless otherwise noted, all lodging is within 30 minutes of the primary airport.

Broadway Inn
264 W. 46th St.
(212) 997-9200
Complimentary Continental breakfast.
Rates: $85–$185

Carlyle
Madison Ave. at E. 76th St.
(212) 744-1600
Grand piano in many suites.
Rates: $300–$2000

Four Seasons
57 E. 57th St.
(212) 758-5700
Web site: http://www.fshr.com
Rates: $440–$7000

Sheraton Manhattan
790 7th Ave.
(212) 581-3300
Rates: $210–$650

The St. Regis Hotel
2 E. 55th St.
(212) 753-4500
Rates: $495–$5000

Waldorf-Astoria
301 Park Ave.
(212) 355-3000
Rates: $300–$1050

Restaurants

Chanterelle
2 Harrison St.
(212) 966-6960
Specializes in contemporary French cuisine with a menu that changes monthly.
Dinner: $75–$89

Daniel
20 E. 76th St.
(212) 288-0033
French menu with seasonal cuisine.
Dinner: $69–$110

Fifty Seven Fifty Seven
Four Seasons Hotel
57 E. 57th St.
(212) 758-5700
Specializes in Maryland crab cakes.
Dinner: $24–$31

Hudson River Club
250 Vesey St.
(212) 786-1500
Specializes in Hudson River Valley dishes.
Dinner: $60

Le Chantilly
106 E. 57th St.
(212) 751-2931
French menu.
Dinner: $55

San Domenico
240 Central Park S.
(212) 265-5959
Specializes in roasted veal loin.
Dinner: $17–$43

Windows on the World
1 World Trade Center
(212) 524-7011
Specializes in seafood.
Dinner: $26–$35

Syracuse

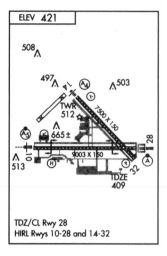

ELEV 421

508 ∧

497 ∧

∧503

TWR 512

665±

∧
513

9003 x 150

TDZE 409

TDZ/CL Rwy 28
HIRL Rwys 10-28 and 14-32

Airport   Syracuse Hancock International (SYR) is located 4 miles northeast of the city. Coordinates: N43° 06.67' W076° 06.38'.

Traffic Pattern   The traffic pattern is flown at 1221 ft MSL. Lefthand pattern for Runways 10, 14, 28, and 32. Runways 10 and 28 are asphalt surfaced, 9003 ft long by 150 ft wide. Runways 14 and 32 are asphalt surfaced, 7500 ft long by 150 ft wide.

FBO   Sair Aviation, Inc. (315) 455-7951. Hours: 24. Frequency 122.95.

Navigational Information   SYR is located on the New York Sectional Chart or L12, L26 low en route chart. From the SYR 117.0 VOR 135° at 5.2 miles.

Instrument Approaches   ILS, VOR, VOR/DME, VOR(TAC), NDB, and GPS approaches are available.

Cautions   Deer, foxes, and birds are in the vicinity.

Fuel   100LL and JetA are available. Exxon and most major credit cards are accepted.

Frequencies
TOWER 120.3   GROUND 121.7   CLEARANCE DELIVERY 125.05
APPROACH SYRACUSE 124.6   UNICOM 122.95   ATIS 132.05.

Runway Lights   Operated by the tower.

Transportation

Rental Cars
Alamo, (315) 454-8890
Avis, (315) 455-9601

Budget, (315) 454-8156
Hertz, (315) 455-2496
National, (315) 455-7495
Thrifty, (315) 455-7012

Taxicabs
Century, (315) 455-5151

Courtesy Cars   None available.

Area Attractions

Burnet Park Zoo.   S. Wilbur Ave. (315) 435-8516. Various exhibits. Admission charged.

Erie Canal Museum.   Erie Blvd. E. and Montgomery St. (315) 471-0593. Exhibits detail the construction of the Erie Canal. Donation.

Everson Museum of Art.   401 Harrison St. (315) 474-6064. Collection of American paintings. Free.

Landmark Theatre.   362 S. Salina St. (315) 475-7979. Built in 1928 in the era of vaudeville-movie houses.

Other Attractions

Lowe Art Gallery.   Shaffer Art Building. (315) 443-3127. Paintings and sculpture. Free.

New York State Canal Cruises.   (800) 545-4318. Sightseeing cruises.

Onondaga Lake Park.   (315) 453-6712. Picnicking. Free.

Interesting Facts & Events

Mid-June.   Balloon Festival.

July.   Onondaga Lake Waterfront Extravaganza. Water-ski show.

Mid-August.   Scottish Games.

Late August–early September. New York State Fair. State Fairgrounds. (315) 487-7711.

For More Information Contact the Convention and Visitor's Bureau, Greater Syracuse Chamber of Commerce at 572 S. Salina St., (315) 470-1800, for more information, a tourist package, and discounts.

Lodging

Unless otherwise noted, all lodging is within 30 minutes of the primary airport.

Courtyard by Marriott
6415 Yorktown Circle
(315) 432-0300
Rates: $79–$99

Embassy Suites
6646 Old Collamer Rd.
(315) 446-3200
Airport transportation.
Rates: $109–$169

Four Points by Sheraton
Electronics Pkwy. and 7th North St.
(315) 457-1122
Rates: $75–$150

Hampton Inn
6605 Old Collamer Rd.
(315) 463-6443
Health club privileges.
Rates: $64–$75

Marriott
6301 NY 298
(315) 432-0200
Free airport transportation.
Rates: $140–$275

Ramada Inn
1305 Buckley Rd.
(315) 457-8670
Airport transportation.
Rates: $70–$150

Restaurants

Coleman's
100 S. Lowell Ave.
(315) 476-1933
Specializes in corned beef.
Dinner: $10–$17

Green Gate Inn
2 Main St.
(315) 672-9276
Specializes in steak.
Dinner: $7–$19

Pascale's
204 W. Fayette St.
(315) 471-3040
Continental menu with wood-burning grill.
Dinner: $10–$20

Ohio

Akron

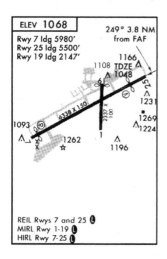

```
ELEV 1068          249° 3.8 NM
Rwy 7 ldg 5980'      from FAF
Rwy 25 ldg 5500'
Rwy 19 ldg 2147'        1166
              1108 TDZE △
                    △ 1048
                        △
                        123ı
              633B X 150      ■1269
1093                      △1224
△_x__x__1262    I   △
        x        ☆  1196


REIL Rwys 7 and 25 ⓵
MIRL Rwy 1-19 ⓵
HIRL Rwy 7-25 ⓵
```

Akron is considered the Rubber Capital of the World and is also home to the National Inventors Hall of Fame.

Airport Akron Fulton International (AKR) is located just south of the city. Coordinates: N41° 02.24' W081° 28.04'.

Traffic Pattern The traffic pattern is flown at 1818 ft MSL. Lefthand pattern for Runways 1, 7, 19, and 25. Runways 1 and 19 are asphalt surfaced, 2337 ft long by 100 ft wide. Runways 7 and 25 are asphalt surfaced, 5998 ft long by 150 ft wide.

FBO Airspect Air, Inc. (330) 794-8383. Hours: 8 A.M.–8 P.M., 24 hr self-service fuel. Frequency 122.95.

Navigational Information AKR is located on the Detroit Sectional Chart or L23, L24 low en route chart. From the ACO 114.4 VOR 255° at 12.8 miles.

Instrument Approaches LOC, NDB, and GPS approaches are available.

Cautions Power lines and hills are south and west.

Fuel 100LL and JetA are available. Most major credit cards are accepted.

Frequencies
CLEARANCE DELIVERY 121.6   APPROACH AKRON–CANTON 118.6
UNICOM/CTAF 123.075.

Runway Lights Tower and pilot-controlled lighting.

Transportation

Rental Cars
Airspect Air, (330) 794-8383, on the field.

Taxicabs
Carey Limousine Service, (330) 253-6743
Yellow, (330) 253-3141

Courtesy Cars The FBO has one available.

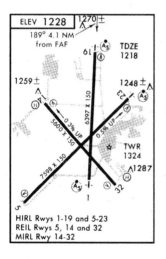

Airport Akron–Canton Regional (CAK) is located 10 miles southeast of the city. Coordinates: N40° 54.98' W081° 26.55'.

Traffic Pattern The traffic pattern is flown at 2300 ft MSL. Lefthand pattern for Runways 1, 5, 14, 19, 23, and 32. Runways 14 and 32 are asphalt surfaced, 5600 ft long by 150 ft wide. Runways 1 and 19 are asphalt surfaced, 6397 ft long by 150 ft wide. Runways 5 and 23 are asphalt surfaced, 7598 ft long by 150 ft wide.

FBO A-Flite, Inc. (330) 494-5560. Hours: 24. Frequency 122.95.

FBO ASW Aviation Services. (216) 494-6104 or (800) 423-5047. Hours: 6 A.M.–10 P.M. Frequency 122.95.

FBO McKinley Air Transport. (330) 499-3316 or (800) 225-6446. Hours: 24. Frequency 122.95.

Navigational Information CAK is located on the Detroit Sectional Chart or L23, L24 low en route chart. From the ACO 114.4 VOR 228° at 15.9 miles.

Instrument Approaches ILS, VOR, ASR, and GPS approaches are available.

Cautions Deer and birds are in the vicinity.

Fuel 100LL and JetA are available. Phillips, Avfuel, Air BP, and most major credit cards are accepted.

Frequencies
TOWER 118.3   GROUND 121.7   CLEARANCE DELIVERY 132.05
APPROACH AKRON 125.5, AKRON 118.6, AKRON 126.4   UNICOM
122.95   ATIS 121.05.

Runway Lights Operated by the tower.

Transportation

Rental Cars Available at the terminal.
Avis, (330) 896-2457
Budget, (330) 253-3540
Hertz, (330) 896-1331
National, (800) 227-7368

Taxicabs
Airport Limo, (330) 928-8172
Shuttle One Service, (330) 494-5800
Zona Express, (330) 497-3738

Courtesy Cars None available.

Area Attractions

Akron Art Museum. 70 E. Market St. (330) 376-9185. Changing exhibits. Free.

Akron Zoological Park. 500 Edgewood Ave. (330) 375-2525. This 26-acre zoo features birds, mammals, and reptiles from around the world. Admission charged.

Goodyear World of Rubber. 1201 E. Market St. (330) 796-6546. Tour includes movies on tire production. Free.

Inventure Place. 221 S. Broadway St. (330) 762-6565. Site of the National Inventors Hall of Fame. Features a wide array of programs, activities, workshops, and exhibits.

Other Attractions

Cuyahoga Valley Scenic Railroad. (800) 468-4070. Journey through Cuyahoga Valley. Admission charged.

Hale Farm and Village. 2686 Oak Hill Rd. (330) 666-3711 or (800) 589-9703. Depicts northeastern Ohio's rural life in the mid-1800s. Craft demonstrations, farming. Admission charged.

John Brown Home. 514 Diagonal Rd. (330) 535-1120. Abolitionist's residence.

Stan Hywet Hall and Gardens. 714 W. Portage Path. (330) 836-5533. Stroll through 70 acres of grounds and gardens. Open daily except Monday. Admission charged.

The Winery at Wolf Creek. 2637 S. Cleveland Massillon Rd. (330) 666-9285. Free.

Interesting Facts & Events

First or second weekend in August. All-American Soap Box Derby. Municipal Airport. (330) 666-9285.

Last three weekends in September. Yankee Peddler Festival. Clay's Park Resort. Arts, crafts, and entertainment. (800) 535-5634.

For More Information Contact the Akron/Summit Convention and Visitor's Bureau at 77 E. Mill St., (330) 374-7560 or (800) 245-4254, for more information, a tourist package, and discounts.

Lodging

Unless otherwise noted, all lodging is within 30 minutes of the primary airport.

Comfort Inn West
130 Montrose West Ave.
(330) 666-5050
Complimentary breakfast.
Rates: $55–$95

Holiday Inn–South
I-77 and Arlington Rd.
(330) 644-7126
Airport transportation.
Rates: $59–$185

Hilton Inn–West
3180 W. Market St.
(330) 867-5000
Airport transportation.
Rates: $79–$140

Radisson Inn
200 Montrose West Ave.
(330) 666-9300
Airport transportation.
Rates: $100–$150

Restaurants

Carousel Dinner Theatre
1275 E. Waterloo Rd.
(330) 724-9855 or
(800) 362-4100
Professional dinner theater
features Broadway musicals.

Lanning's
826 N. Cleveland–Massillon Rd.
(330) 666-1159
Specializes in steak.
$14–$28

Mill Street Tavern
135 S. Broadway
(330) 762-9333
Specializes in prime rib.
Dinner: $15–$20

Cincinnati

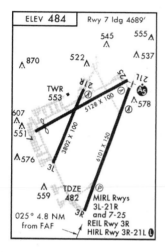

Nicknamed the Queen City, Cincinnati is located in the southwestern part of the state. The Ohio River bordering the city provides numerous recreational activities for residents and visitors alike.

Airport Cincinnati Municipal–Lunken Field (LUK) is located 3 miles southeast of the city. Coordinates: N39° 06.20' W084° 25.12'.

Traffic Pattern The traffic pattern is flown at 1480 ft MSL. Lefthand pattern for Runways 3L, 3R, 7, 21L, 21R, and 25. Runways 3L and 21R are asphalt surfaced, 3802 ft long by 100 ft wide. Runways 3R and 21L are asphalt surfaced, 6102 ft long by 150 ft wide. Runways 7 and 25 are asphalt surfaced, 5128 ft long by 100 ft wide.

FBO Midwest Jet Center. (513) 871-8600. Hours: 24. Frequency 122.95.

FBO Million Air Cincinnati. (513) 871-2020. Hours: 24. Frequency 122.95.

Navigational Information LUK is located on the Cincinnati Sectional Chart or L22 low en route chart. From the CVG 117.3 VOR 068° at 14.3 miles.

Instrument Approaches ILS, LOC BC, NDB, and GPS approaches are available.

Fuel 100LL and JetA are available. Air BP and most major credit cards are accepted.

Frequencies

TOWER 118.7, 133.925   GROUND 121.9   CLEARANCE DELIVERY
121.9   APPROACH CINCINNATI 121.0   UNICOM 122.95   CTAF
118.7   ATIS 120.25.

Runway Lights   Tower and pilot-controlled lighting.

Transportation

Rental Cars

Enterprise, (513) 871-2020 or (513) 871-8600

Taxicabs

Carey Limousine Service, (513) 531-7321
Dispatch, (513) 241-2100
East, (513) 231-8877

Courtesy Cars   Both FBOs have one available.

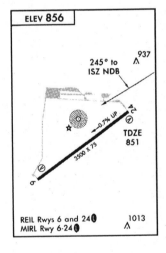

Airport   Cincinnati–Blue Ash (ISZ) is located 6 miles northeast of the city. Coordinates: N39° 14.80' W084° 23.34'.

Traffic Pattern   The traffic pattern is flown at 1700 ft MSL. Lefthand pattern for Runway 6. Righthand pattern for Runway 24. Runways 6 and 24 are asphalt surfaced, 3500 ft long by 75 ft wide.

FBO   Co-op Aircraft Service, Inc. (513) 791-8500. Hours: 6:30 A.M.–dark weekdays, 8 A.M.–dark weekends. Frequency 123.0.

FBO   Exec Aviation, Inc. (513) 984-3881. Hours: 7:30 A.M.–6:30 P.M. Frequency 123.0.

Navigational Information ISZ is located on the Cincinnati Sectional Chart or L22 low en route chart. From the CVG 117.3 VOR 047° at 20.2 miles.

Instrument Approaches VOR, NDB, and GPS approaches are available.

Cautions Power lines are north and deer are in the vicinity.

Fuel 100LL and JetA are available. Air BP and most major credit cards are accepted.

Frequencies
CLEARANCE DELIVERY 124.9 APPROACH CINCINNATI 121.0 UNICOM/CTAF 123.0.

Runway Lights Tower and pilot-controlled lighting.

Transportation

Rental Cars
Available at Co-op Aircraft Service, (513) 791-8500.
Exec Aviation, Inc. (513) 984-3881

Taxicabs
Elite, (513) 733-8294
Tatman, (513) 812-2068

Courtesy Cars None available.

Area Attractions

Cincinnati Art Museum. (513) 721-5204. Paintings, sculpture, photographs, and decorative arts. Admission charged.

Cincinnati Fire Museum. 315 W. Court St. (513) 621-5553. Restored firehouse with exhibits. Admission charged.

Cincinnati History Museum. 1301 Western Ave. (513) 287-7030. Permanent and temporary exhibits. Admission charged.

Children's Museum of Cincinnati. 700 W. Pete Rose Way. (513) 521-5437. Features 200 hands-on displays for preschoolers. Admission charged.

Cincinnati Zoo and Botanical Garden. 3400 Vine St. (513) 281-4700. There are 750 species in a variety of habitats. Admission charged.

Harriet Beecher Stowe Memorial. 2950 Gilbert Ave. (513) 632-5120. The author of Uncle Tom's Cabin lived here from 1832 to 1836. Donation.

Museum of National History and Science. 1301 Western Ave. (513) 287-7020. Natural history of Ohio Valley. Admission charged.

William Howard Taft National Historic Site. 2038 Auburn Ave. (513) 684-3262. Boyhood home of the twenty-seventh president. Free.

Other Attractions

Chateau Laroche. (513) 683-4686. The only medieval castle in the United States. Admission charged.

Cincinnati Bengals. Cinergy Field. (513) 621-3550. Professional football.

Cincinnati Reds. Cinergy Field. (513) 421-4510. Major League baseball.

River Downs Race Track. 6301 Kellogg Ave. (513) 232-8000. Thoroughbred racing.

Interesting Facts & Events

Last two weekends in May. May Festival. Choral festival. (513) 381-3300.

Late July. Coors Light Festival. Performers in rhythm and blues. (513) 871-3900.

Labor Day weekend. Riverfest. Entertainment includes performances, amusement rides, and fireworks.

Mid-September. Oktoberfest-Zinzinnati. German music, singing, dancing, and of course, beer.

For More Information   Contact the Greater Cincinnati Convention and Visitor's Bureau at 300 W. 6th St., (800) CINCY-USA, for more information, a tourist package, and discounts.

Lodging

Unless otherwise noted, all lodging is within 30 minutes of the primary airport.

Cincinnatian
600 Vine St.
(513) 381-3000
Rates: $250–$1500

Courtyard by Marriott
4625 Lake Forest Dr.
(513) 733-4334
Rates: $94–$119

Embassy Suites
4554 Lake Forest Dr.
(513) 733-8900
E-mail: cybga@aol.com
Complimentary full breakfast.
Rates: $89–$179

Hampton Inn
10900 Crowne Point Dr.
(513) 771-6888
Complimentary Continental breakfast.
Rates: $79–$89

Holiday Inn
4501 East Gate Blvd.
(513) 752-4400
Business center.
Rates: $94–$275

Westin
Fountain Square
5th and Vine Sts.
(513) 621-7700
Rates: $300–$1200

Restaurants

Celestial
1071 Celestial St.
(513) 241-4455
Specializes in fresh seafood.
Dinner: $19–$25

Heritage
7664 Wooster Pike
(513) 561-9300
e-mail: theheritage.com
Specializes in American cuisine.
Dinner: $14–$21

Jungle Jim's International Market
5440 Dixie Hwy.
(513) 829-1919
With 7000 wines and over 4 acres
of food from around the world.
Prices vary.

Maisonette
114 E. 6th St.
(513) 721-2260
Exceptional French cuisine.
Dinner: $25–$36

Cleveland

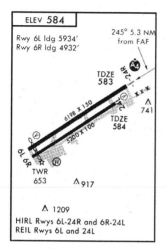

ELEV 584

Rwy 6L ldg 5934'
Rwy 6R ldg 4932'

245° 5.3 NM
from FAF

TDZE
583

TDZE
584

741

TWR
653

917

1209

HIRL Rwys 6L-24R and 6R-24L
REIL Rwys 6L and 24L

6198 X 150

5200 X 100

No longer the mistake on the lake, Cleveland is a vibrant bustling city. It's home to the Rock 'n' Roll Hall of Fame and numerous other attractions.

Airport Burke Lakefront (BKL) is located 1 mile north of the city. Coordinates: N41° 31.05' W081° 41.01'.

Traffic Pattern The traffic pattern is flown at 1550 ft MSL. Lefthand pattern for Runway 6R. Righthand pattern for Runway 24L. Runways 6R and 24L are asphalt surfaced, 5200 ft long by 100 ft wide.

FBO Business Aircraft Center. (216) 781-1200. Hours: 24. Frequency 122.95. ·

FBO Million Air Cleveland. (216) 861-2030. Hours: 24. Frequency 122.95.

Navigational Information BKL is located on the Detroit Sectional Chart or L23, L24 low en route chart. From the CXR 112.7 VOR 275° at 23.4 miles.

Instrument Approaches LOC, NDB, and GPS approaches are available.

Cautions Sea gulls are in the vicinity.

Fuel 100LL and JetA are available. Avfuel, Air BP, and most major credit cards are accepted.

Frequencies
TOWER 124.3   GROUND 121.9   CLEARANCE DELIVERY 121.9
APPROACH CLEVELAND 125.35   UNICOM 122.95   CTAF 124.3
ATIS 125.25.

Runway Lights Operated by the tower.

Transportation

Rental Cars
Budget, (216) 574-9871

Taxicabs
Americab, (216) 881-1111
Company C.A.R., (216) 861-7433 or (800) 413-2527
Yellow, (216) 623-1500, direct line on the field.

Courtesy Cars Both FBOs have one available.

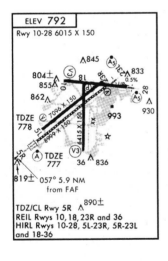

Airport Cleveland–Hopkins International (CLE) is located 9 miles southwest of the city. Coordinates: N41° 24.65' W081° 50.96'.

Traffic Pattern The traffic pattern is flown at 1800 ft MSL. Lefthand pattern for Runways 5L, 5R, 10, 18, 23L, 23R, 28, and 36. Runways 5L and 23R are concrete surfaced, 7095 ft long by 150 ft wide. Runways 5R and 23L are concrete surfaced, 8999 ft long by 150 ft wide. Runways 10 and 28 are asphalt surfaced, 6015 ft long by 150 ft wide. Runways 18 and 36 are asphalt surfaced, 6415 ft long by 150 ft wide.

FBO I-X Jet Center. (216) 362-1500 or (800) 688-3314. Hours: 24. Frequency 122.95.

Navigational Information CLE is located on the Detroit Sectional Chart or L23 low en route chart. From the DJB 113.6 VOR 082° at 14.5 miles.

Instrument Approaches ILS, NDB, VOR/DME, and GPS approaches are available.

Fuel 100LL and JetA are available. Texaco and most major credit cards are accepted.

Frequencies
TOWER 120.9, 124.5   GROUND 121.7   CLEARANCE DELIVERY 125.05   DEPARTURE 118.9   APPROACH CLEVELAND 126.55, CLEVELAND 124.0, CLEVELAND 123.85   UNICOM 122.95   ATIS 127.85.

Runway Lights Operated by the tower.

Transportation

Rental Cars
Avis, (216) 265-3700
Budget, (216) 433-4433
Dollar, (216) 267-3133
Hertz, (216) 267-8900
National, (216) 267-0060

Taxicabs
Americab, (216) 881-1111
Yellow Cab, (216) 623-1500

Courtesy Cars The FBO has one available.

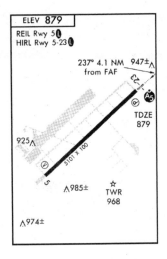

Airport Cuyahoga County (CGF) is located 10 miles east of the city. Coordinates: N41° 33.90' W081° 29.19'.

Traffic Pattern The traffic pattern is flown at 1900 ft MSL. Lefthand pattern for Runways 5 and 23. Runways 5 and 23 are asphalt surfaced, 5101 ft long by 100 ft wide.

FBO Beckett Enterprises, Inc. (216) 261-1111. Hours: 24. Frequency 122.95.

FBO Corporate Wings. (216) 261-5900. Hours: 24. Frequency 122.95.

Navigational Information CGF is located on the Detroit Sectional Chart or L23 low en route chart. From the CXR 112.7 VOR 286° at 14.8 miles.

Instrument Approaches ILS, LOC BC, NDB, and GPS approaches are available.

Cautions Birds and deer are in the vicinity.

Fuel 100LL and JetA are available. Texaco, Phillips, and most major credit cards are accepted.

Frequencies
TOWER 118.5   GROUND 121.85   CLEARANCE DELIVERY 121.8
APPROACH CLEVELAND 125.35   UNICOM 122.95   CTAF 118.5.

Runway Lights Tower and pilot-controlled lighting.

Transportation

Rental Cars
Grand, (216) 261-1111 or (216) 261-5900

Taxicabs
Americab, (216) 881-1111
Yellow, (216) 623-1500

Courtesy Cars   Both FBOs have one available.

Area Attractions

Cleveland Botanical Gardens.   11030 East Blvd. (216) 721-1600. There are 10 acres of landscaped gardens. Free.

Cleveland Hungarian Heritage Museum.   Richmond Mall at Richmond and Wilson Mills Rds. (216) 605-1574. The museum showcases Hungarian traditions and culture. Free.

Cleveland Metroparks Zoo.   3900 Brookside Park Dr. (216) 661-6500. Seventh oldest zoo in the country with 3300 animals. Admission charged.

The Cleveland Museum of Art.   11150 East Blvd. at University Circle. (216) 421-7340. The museum's 30,000 works of art represent a wide range of history and culture. Free.

Cleveland Museum of Natural History.   Wade Oval at University Circle. (216) 231-4600. Dinosaurs, birds, gems, and exhibits of Ohio. Admission charged.

NASA Lewis Visitor Center.   21000 Brookpark Rd. (216) 433-2001. Display features space station, Skylab 3, an Apollo capsule, and communications satellites. Free.

Rock 'n' Roll Hall of Fame and Museum.   E. 9th St. Pier. (216) 781-ROCK. Explore rock's impact on culture in this 50,000 ft^2 museum. Admission charged.

The Western Reserve Historical Society.   10825 East Blvd. (216) 721-5722. Nearly 200 automobiles and aircraft. Admission charged.

Other Attractions

Steamship William G. Mather Museum. 1001 E. 9th St. Pier. (216) 574-6262. Built in 1925 to carry iron ore, it now exhibits the heritage of the "iron boats." Admission charged.

U.S.S. COD. N. Marginal Rd. (216) 566-8770. Tours. Admission charged.

Interesting Facts & Events

March. Tri-City JassFest. (216) 987-4444.

2nd Saturday in June. Clifton Arts and MusicFest. Free annual family event. (216) 228-4383.

Labor Day weekend. Cleveland National Air Show. Burke Lakefront Airport. (216) 781-0747.

For More Information Contact the Convention and Visitor's Bureau of Greater Cleveland, 3100 Terminal Tower, (800) 321-1001, for more information, a tourist package, and discounts.

Lodging

Unless otherwise noted, all lodging is within 30 minutes of the primary airport.

Comfort Inn
17550 Rosbough Dr.
(216) 234-3131
Free airport transportation.
Rates: $79–$99

Marriott Airport
4277 W. 150th St.
(216) 252-5333
Free airport transportation.
Rates: $175–$350

Radisson Inn–Cleveland Airport
25070 Country Club Blvd.
(216) 734-5060
Free airport transportation.
Rates: $139–$160

Renaissance
24 Public Square
(216) 696-5600
Health club privileges.
Rates: $125–$1500

The Ritz–Carlton
1515 W. 3d St.
(216) 623-1300
Rates: $149–$359

Bed & Breakfast

Baricelli
2203 Cornell Rd
(216) 791-6500
Individually decorated rooms.
Rates: $125–$150

Restaurants

Café Sausalito
The Galleria at Erieview
1301 E. 9th St.
(216) 696-2233
Specializes in seafood.
Dinner: $15–$25

John Q's Steakhouse
55 Public Square
(216) 861-0900
Cleveland's best steakhouse.
Dinner: $20–$25

Pier W Restaurant
12700 Lake Ave. at the Winton Pl.
(216) 228-2250
Specializes in seafood.
Dinner: $20–$25

Watermark Restaurant
1250 Old River Rd.
(216) 241-1600
Specializes in seafood.
Dinner: $15–$20

Dayton

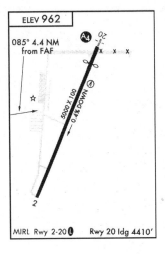

ELEV 962

085° 4.4 NM
from FAF

5000 X 100
0.4% DOWN

2

MIRL Rwy 2-20 | Rwy 20 ldg 4410'

Dayton is the birthplace of aviation. It was here that the Wright Brothers began their experiments that led to powered flight.

Airport Dayton–Wright Brothers (MGY) is located 10 miles south of the city. Coordinates: N39° 35.47' W084° 13.43'.

Traffic Pattern The traffic pattern is flown at 1960 ft MSL. Lefthand pattern for Runways 2 and 20. Runways 2 and 20 are asphalt surfaced, 4981 ft long by 100 ft wide.

FBO Aviation Sales, Inc. South. (513) 885-3662. Hours: 8 A.M.–8 P.M. Frequency 122.8.

FBO Pax Air, Inc. (513) 885-1434. Hours: 8 A.M.–8 P.M., after hours on request. (513) 433-0581. Frequency 122.95.

Navigational Information MGY is located on the Cincinnati Sectional Chart or L23 low en route chart. From the FFO 115.2 VOR 214° at 15.7 miles.

Instrument Approaches LOC, NDB, and GPS approaches are available.

Fuel 100LL and JetA are available. Exxon, Avfuel, and most major credit cards are accepted.

Frequencies
CLEARANCE DELIVERY 119.4   DEPARTURE 126.5   APPROACH DAYTON 118.85   UNICOM/CTAF 122.8.

Runway Lights Tower and pilot-controlled lighting.

Transportation

Rental Cars

Enterprise, (513) 439-9001

Taxicabs

Yellow, (513) 228-1155

Courtesy Cars None available.

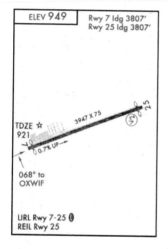

ELEV 949

Rwy 7 ldg 3807'
Rwy 25 ldg 3807'

3947 X 75

TDZE ☆
921

0.7% UP

068° to
OXWIF

URL Rwy 7-25
REIL Rwy 25

Airport Greene County (Lewis A. Jackson Regional) (I19) is located 10 miles east of the city. Coordinates: N39° 41.50' W083° 59.41'.

Traffic Pattern The traffic pattern is flown at 2000 ft MSL. Lefthand pattern for Runways 7 and 25. Runways 7 and 25 are asphalt surfaced, 3947 ft long by 75 ft wide.

FBO Commander-Aero, Inc. (513) 376-2925. Hours: 8 A.M.–6 P.M. Frequency 122.7.

Navigational Information I19 is located on the Cincinnati Sectional Chart or L23 low en route chart. From the MXQ 112.9 VOR 335° at 18 miles.

Instrument Approaches An NDB approach is available.

Cautions Deer are in the vicinity.

Fuel 100LL and JetA are available. Avfuel and most major credit cards are accepted.

Frequencies

DEPARTURE   118.85   APPROACH   DAYTON   118.85
UNICOM/CTAF 122.7.

Runway Lights Tower and pilot-controlled lighting.

Transportation

Rental Cars

Alen Besco, (513) 372-6074

Eastgate Ford, (513) 429-1300

Enterprise, (513) 427-0100

Taxicabs

Xenia Cab Co., (513) 372-1441

Courtesy Cars The FBO has one available.

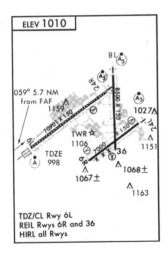

Airport James M. Cox Dayton International (DAY) is located 9 miles north of the city. Coordinates: N39° 54.14' W084° 13.16'.

Traffic Pattern The traffic pattern is flown at 1800 ft MSL. Lefthand pattern for Runways 6L, 6R, 18, 24L, 24R, and 36. Runways 6L and 24R are asphalt surfaced, 10,900 ft long by 150 ft wide. Runways 6R and 24L are asphalt surfaced, 7000 ft long by 150 ft wide. Runways 18 and 36 are asphalt surfaced, 8500 ft long by 150 ft wide.

FBO Aviation Sales. (513) 898-3927. Hours: 6 A.M.–8 P.M. weekdays. Frequency 122.95.

FBO Stevens Aviation. (513) 454-3400 or (800) 359-7232. Hours: 24. Frequency 122.95.

FBO Wright Brothers Aero. (513) 890-8900. Hours: 6 A.M.–11 P.M. daily, 24 hours on request. Frequency 122.95.

Navigational Information DAY is located on the Cincinnati Sectional Chart or L23 low en route chart. From the DQN 114.5 VOR 131° at 10.7 miles.

Instrument Approaches ILS, NDB, ASR, VOR/DME, and GPS approaches are available.

Fuel 100LL and JetA are available. Exxon, Air BP, and most major credit cards are accepted.

Frequencies
TOWER 119.9   GROUND 121.9   CLEARANCE DELIVERY 121.75
APPROACH DAYTON 134.45, DAYTON 118.85, DAYTON 127.65
UNICOM 122.95   ATIS 125.8.

Runway Lights Operated by the tower.

Transportation

Rental Cars
Alamo, (513) 454-8450
Avis, (513) 898-8357
Budget, (513) 898-9321
Dollar, (513) 454-8430
Hertz, (513) 898-5806
National, (513) 890-0100

Taxicabs
Charter Vans, (513) 898-4043
Yellow Cab, (513) 228-1155

Courtesy Cars A courtesy car is available at both FBOs.

Area Attractions

The Dayton Art Institute. 456 Bemonte Park N. (937) 223-5277 or (800) 296-4426. One of the nation's finest art museums. Free.

Dayton Museum of Discovery. 2600 DeWeese Pkwy. (937) 275-7431. Natural history exhibits and Philips Space Theater shows. Admission charged.

SunWatch Prehistoric Indian Village. 2301 W. River Rd. (937) 268-8199. Reconstructed 12th-century houses and gardens. Admission charged.

United States Air Force Museum. Springfield and Harshman Rds. (937) 255-3284. Comprehensive military aviation exhibits and IMAX theater. Free.

Other Attractions

Wright Brothers Memorial. OH 444 at Jct. Old OH 4.

Wright State University. E 3d St. (937) 873-2310. Largest collections of Wright Brothers memorabilia.

Interesting Facts & Events

Late July. U.S. Air and Trade Show. (937) 898-5901.

Labor Day weekend. Montgomery County Fair. 1043 S. Main St. (937) 224-1619.

For More Information Contact the Dayton/Montgomery County Convention and Visitor's Bureau at Chamber Plaza, (937) 226-8211, for more information, a tourist package, and discounts.

Lodging

Unless otherwise noted, all lodging is within 30 minutes of the primary airport.

Fairfield Inn by Marriott
6960 Miller Ln.
(937) 898-1120
Rates: $58–$66

Hampton Inn
8099 Old Yankee St.
(937) 436-3700
Rates: $55–$65

Holiday Inn I-675 Conference
Center
2800 Presidential Dr.
(937) 426-7800
Free airport transportation.
Rates: $69–$125

Marriott
1414 S. Patterson Blvd.
(937) 223-7853
Rates: $120–$400

Bed & Breakfast

Lakewood Farm
8495 OH 48
(937) 885-9850
Country atmosphere.
Rates: $65–$75

Restaurants

King Cole
40 N. Main St. at 2d St.
(937) 222-6771
Specializes in baby rack of lamb.
Dinner: $17–$25

L'Auberge
4120 Far Hills Ave.
(937) 299-5536
Specializes in imported fresh
seafood and game.
Dinner: $22–$28

Peasant Stock
424 E. Stroop Rd.
(937) 293-3900
Specializes in fresh fish.
Dinner: $11–$19

Toledo

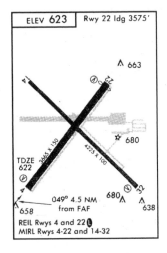

ELEV 623   Rwy 22 ldg 3575'

Λ 663

☆ 680

TDZE
622

049° 4.5 NM   680Λ
from FAF

658   638

REIL Rwys 4 and 22
MIRL Rwys 4-22 and 14-32

Airport Metcalf Field (TDZ) is located 6 miles southeast of the city. Coordinates: N41° 33.84' W083° 28.87'.

Traffic Pattern The traffic pattern is flown at 1500 ft MSL, all aircraft. Lefthand pattern for Runways 4, 14, 22, and 32. Runways 14 and 32 are asphalt surfaced, 4225 ft long by 100 ft wide. Runways 4 and 22 are asphalt surfaced, 3665 ft long by 150 ft wide.

FBO Crow Executive Air, Inc. (419) 838-6921. Hours: 7 A.M. Monday–7 P.M. Friday, 7 A.M.– 7 P.M. weekends. Frequency 122.7.

Navigational Information TDZ is located on the Detroit Sectional Chart or L23 low en route chart. From the VWV 113.1 VOR 048° at 9.8 miles.

Instrument Approaches VOR, VOR/DME, and GPS approaches are available.

Cautions Birds are in the vicinity.

Fuel 100LL and JetA are available. Phillips and most major credit cards are accepted.

Frequencies
CLEARANCE DELIVERY 125.6 APPROACH TOLEDO 126.1 UNICOM/CTAF 122.7.

Runway Lights Operated by the tower.

Transportation

Rental Cars Prior request through FBO.

Taxicabs
Eagle, (419) 255-2323
Express, (419) 351-4497

Courtesy Cars The FBO has one available.

Airport Toledo Express (TOL) is located 10 miles west of the city. Coordinates: N41° 35.21' W083° 48.47'.

Traffic Pattern The traffic pattern is flown at 1480 ft MSL. Lefthand pattern for Runways 7, 16, 25, and 34. Runways 16 and 34 are asphalt surfaced, 5598 ft long by 150 ft wide. Runways 7 and 25 are asphalt surfaced, 10,600 ft long by 150 ft wide.

FBO National Flight Services, Inc. (419) 865-2311. Hours: 24. Frequency 122.95.

FBO TOL Aviation, Inc. (419) 866-9375. Hours: 5 A.M.–11 P.M. daily. Frequency 122.95

Navigational Information TOL is located on the Detroit Sectional Chart or L23 low en route chart. From the VWV 113.1 VOR 319° at 11.1 miles.

Instrument Approaches ILS, VOR/DME, NDB, ASR, and GPS approaches are available.

Cautions Deer are in the vicinity.

Fuel 100LL and JetA are available. Air BP, Sunoco, and most major credit cards are accepted.

Frequencies
TOWER 118.1   GROUND 121.9   CLEARANCE DELIVERY 121.9
APPROACH TOLEDO 134.35, TOLEDO 123.975   UNICOM 122.95
ATIS 118.75.

Runway Lights Operated by the tower.

Transportation

Rental Cars
Avis, (419) 865-5541
Budget, (419) 865-8825
Hertz, (419) 891-1284
National, (419) 865-5512

Taxicabs
Airline, (419) 893-3245
Black & White, (419) 478-8300
Checker, (419) 243-2537
Personal, (419) 882-6363

Courtesy Car Both FBOs have one available.

Area Attractions

S.S. Willis B. Boyer Museum Ship.   26 Main St. (419) 936-3070. This freighter has been restored and houses nautical artifacts. Admission charged.

The Toledo Museum of Art.   2445 Monroe St. at Scottwood. (419) 255-8000. One of the finest art museums in the country. Free.

The Toledo Zoo.   2700 Broadway. (419) 385-5721 or (419) 385-4040. Large mammal collection and a children's zoo. Admission charged.

Other Attractions

Horse Racing. 5700 Telegraph Rd. (419) 476-7751. Harness racing.

Interesting Facts & Events

June 27–28. Annual Crosby Festival of the Arts. Ohio's oldest outdoor art show. (419) 936-2986.

First weekend in August. Northwest Ohio Rib-Off. Restaurants compete in rib cooking.

For More Information Contact the Greater Toledo Convention and Visitor's Bureau at 401 Jefferson, (419) 321-6404 or (800) 243-4667, for more information, a tourist package, and discounts.

Lodging

Unless otherwise noted, all lodging is within 30 minutes of the primary airport.

Clarion–Westgate
3536 Secor Rd.
(419) 535-7070
Rates: $80–$275

Crown Inn
1727 W. Alexis Rd.
(419) 473-1485
Rates: $45–$110

Comfort Inn
3560 Secor Rd.
(419) 531-2666
Rates: $60–$65

Holiday Inn
2340 S. Reynolds Rd.
(419) 865-1361
Rates: $99–$120

Restaurants

Mancy's
953 Phillips Ave.
(419) 476-4154
Web site:
http://www.mancys.com
Specializes in steak.
Dinner: $11–$23

Maumee Bay Brewing Company
27 Broadway St.
(419) 241-1ALE
Toledo's original brew, offers specialty sandwiches.
Entrées: $6–$10

Pennsylvania

Erie

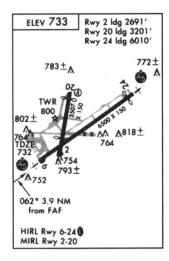

ELEV 733	Rwy 2 ldg 2691' Rwy 20 ldg 3201' Rwy 24 ldg 6010'

Erie is the third largest city and only Great Lakes port in Pennsylvania. It has a fine natural harbor on Lake Erie and is also one of the nation's important industrial centers. The lake and the city take their name from the Native American Eriez tribe.

Airport Erie International (ERI) is located 5 miles southwest of the city. Coordinates: N42° 04.92' W080° 10.57'.

Traffic Pattern The traffic pattern is flown at 1530 ft MSL. Lefthand pattern for Runways 2, 6, 20, and 24. Runways 2 and 20 are asphalt surfaced, 3507 ft long by 150 ft wide. Runways 6 and 24 are asphalt surfaced, 6500 ft long by 150 ft wide.

FBO Erie Airways. (814) 833-1188. Hours: 24. Frequency 122.95.

Navigational Information ERI is located on the Detroit Sectional Chart or L12, L23 low en route chart. From ERI 109.4 VOR 059° at 6.5 miles.

Instrument Approaches ILS, VOR, VOR/DME, NDB, ASR, and GPS approaches are available.

Cautions Birds and deer are in the vicinity.

Fuel 100LL and JetA are available. Most major credit cards are accepted.

Frequencies
TOWER 118.1   GROUND 121.9   CLEARANCE DELIVERY 126.8
APPROACH ERIE 121.0, CLEVELAND CENTER 132.4   UNICOM
122.95   CTAF 118.1   ATIS 120.35.

Runway Lights   Tower and pilot-controlled lighting.

Transportation

Rental Cars
Avis, (814) 833-9879
Budget, (814) 838-4502
National, (814) 838-3041

Taxicabs
Erie, (814) 455-4441

Courtesy Cars   The FBO has one available.

Area Attractions

Erie Art Museum.   411 State St. (814) 459-5477. Art exhibits in a variety of media. Classes, concerts, lectures, and workshops are offered. Admission charged.

Erie Historical Museum and Planetarium.   356 W. 6th St. (814) 871-5790. An 1890 Victorian mansion featuring regional history and decorative art exhibits. Admission charged.

Firefighters Historical Museum.   428 Chestnut St. (814) 456-5969. Over 1300 items of firefighting memorabilia displayed with exhibits dating from 1823. The Hay Loft Theatre offers fire safety films. Admission charged.

Other Attractions

Erie Zoo.   3 miles N. of I-90, State St. Exit 7. (814) 864-4091. More than 300 animals including a children's zoo. Admission charged.

Presque Isle State Park. N. of PA 832. (814) 833-7424. There are 3200 acres of swimming, fishing, boating, hiking, cross-country skiing, picnicking, ice-skating, fishing, and a whole lot more.

For More Information Contact the Erie Chamber of Commerce at 1006 State St., (814) 454-7191, for more information, a tourist package, and discounts.

Lodging

Unless otherwise noted, all lodging is within 30 minutes of the primary airport.

Glass House Inn
3202 W. 26th St.
(814) 833-7751 or
(800) 956-7222
Rates: $53–$89

Hampton Inn
3041 W. 12th St.
(814) 835-4200
Rates: $65–$89

Holiday Inn South
18 W. 18th St.
(814) 456-2961
Rates: $70–$90

Ramada Inn
6101 Wattsburg Rd.
(814) 825-3100
Rates: $65–$95

Restaurants

Pufferbelly
414 French St.
(814) 454-1557
Specializes in regional cuisine,
steak, and seafood.
Dinner: $9–$16

Harrisburg

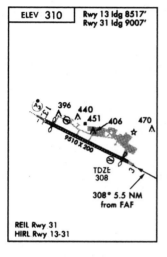

ELEV 310	Rwy 13 ldg 8517' Rwy 31 ldg 9007'

396 440
451 406 470

9510 X 200

TDZE
308

308° 5.5 NM
from FAF

REIL Rwy 31
HIRL Rwy 13-31

Harrisburg is considered to have the finest capitol building in the nation. Its first cornerstone was laid in 1819. John Harris, the area's first settler, opened his trading post here, and his son established the town in 1785.

Airport Harrisburg International (MDT) is located 8 miles southeast of the city. Coordinates: N40° 11.61' W076° 45.80'.

Traffic Pattern The traffic pattern is flown at 1300 ft MSL. Lefthand pattern for Runways 13 and 31. Runways 13 and 31 are concrete surfaced, 9501 ft long by 200 ft wide.

FBO Stambaugh's Air Service. (717) 944-1787. Hours: 5 A.M.–midnight. Frequency 127.2.

Navigational Information MDT is located on the New York Sectional Chart or L24, L28 low en route chart. From the LRP 117.3 VOR 291° at 22.2 miles.

Instrument Approaches ILS, VOR, and GPS approaches are available.

Cautions Birds are in the vicinity.

Fuel 100LL and JetA are available. Air BP and most major credit cards are accepted.

Frequencies
TOWER 124.8   GROUND 121.7   CLEARANCE DELIVERY 127.85
APPROACH HARRISBURG 124.1, LANCASTER 126.45   UNICOM
122.95   ATIS 118.8.

Runway Lights   Operated by the tower.

Transportation

Rental Cars
Avis, (717) 948-3720
Hertz, (717) 944-4088
National, (717) 948-3710
Thrifty, (717) 848-5710

Taxicabs
Aero Corp, (717) 944-4019
Diamonds, (717) 939-7805
Lancaster Limo, (717) 732-3533
Yellow, (717) 238-7252

Courtesy Cars   None available.

Area Attractions

Capitol Hill Buildings.   N. 3d and Walnut Sts. (717) 787-6810. A cluster of buildings in a 45-acre complex including The Capitol, North Office Building, South Office Building, Forum Building, Finance Building, and The State Museum of Pennsylvania. Free.

Indian Echo Caverns.   10 miles E. on U.S. 322. (717) 566-8131. View stalagmite and stalactite formations. Picnic and playground area. Admission charged.

John Harris Mansion.   219 S. Front St. (717) 233-3462. This stone house has 19th-century furnishings, a library, and a collection of county artifacts. Once home of the city's founder, it now serves as Historical Society of Dauphin County headquarters. Admission charged.

Museum of Scientific Discovery.   3d and Walnut Sts., 1st level, Strawberry Square. (717) 233-7969. Scientific and math principles with hands-on exhibits. Admission charged.

Other Attractions

Dauphin County Courthouse. Front and Market Sts. (717) 255-2741. Map on floor of main foyer depicts township boundaries.

Riverfront Park. Relax in the park flanking Front St. with a concrete walk along the Susquehanna River.

Interesting Facts & Events

Early-mid-February. Eastern Sports and Outdoor Show. State Farm Show Building. (717) 232-1377.

Labor Day. Kipona. Boating and water fun activities. (717) 232-4121.

Mid-October. Pennsylvania National Horse Show. State Farm Show Building. (717) 232-4121.

For More Information Contact the Capital Region Chamber of Commerce at 114 Walnut St., (717) 232-4121, or the Harrisburg–Hershey–Carlisle Tourism and Convention Bureau, (717) 232-1377, for more information, a tourist package, and discounts.

Lodging

Unless otherwise noted, all lodging is within 30 minutes of the primary airport.

Best Western
300 N. Mountain Rd.
(717) 652-7180
Rates: $50–$85

Best Western Crown Park
765 Eisenhower Blvd.
(717) 558-9500
Rates: $79–$150

Hampton Inn
4230 Union Deposit Rd.
(717) 545-9595
Rates: $68–$86

Holiday Inn East
4751 Linkle Rd.
(717) 939-7841
Rates: $109–$151

Marriott
4650 Lindle Rd.
(717) 564-5511
Rates: $80–$265

Ramada Hotel–On Market Square
23 S. 2d St.
(717) 234-5021
Rates: $75–$145

Restaurants

Alfred's Victorian Restaurant
38 N. Union St., Middletown
(717) 944-5373
Specializes in homemade pasta.
Dinner: $10–$28

Sal's Bistro
3745 N. 6th St.
(717) 232-4131
Specializes in veal.
Dinner: $8–$16

Manada Hill Inn
128 N. Hershey Rd.
(717) 652-0400
Specializes in seafood and
prime rib.
Dinner: $8–$20

Philadelphia

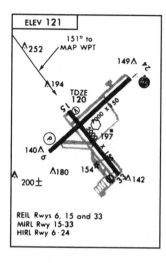

```
ELEV 121
        151° to
 252   /MAP WPT
                149
  194
     TDZE
     120
140
    180   154
200±         142

REIL Rwys 6, 15 and 33
MIRL Rwy 15-33
HIRL Rwy 6-24
```

Philadelphia was the nation's first capital. It is here that the Declaration of Independence was written and adopted, the Constitution was molded and signed, the Liberty Bell was rung, and Washington served most of his years as president. Philadelphia is known as the City of Brotherly Love. It is the second largest city on the East Coast and has served as an arsenal and a shipyard during four wars.

Airport   Northwest Philadelphia (PNE) is located 10 miles northeast of the city. Coordinates: N40° 04.92' W075° 00.64'.

Traffic Pattern   The traffic pattern is flown at 1621 ft MSL. Lefthand pattern for Runways 6, 15, 24, and 33. Runways 15 and 33 are asphalt surfaced, 5000 ft long by 150 ft wide. Runways 6 and 24 are asphalt surfaced, 7000 ft long by 150 ft wide.

FBO   Delaware Aviation. (215) 698-3100. Hours: 24. Frequency 122.95.

FBO   North Philadelphia Jet Center. (215) 673-9000. Hours: 7 A.M.–11 P.M. Frequency 122.95.

Navigational Information   PNE is located on the New York Sectional Chart or L24, L28 low en route chart. From the ARD 108.2 VOR 215° at 11.3 miles.

Instrument Approaches   ILS, LOC BC, VOR, VOR/DME, and GPS approaches are available.

Cautions   Deer and birds are in the vicinity.

Fuel 100LL and JetA are available. Texaco, Exxon, and most major credit cards are accepted.

Frequencies
TOWER 126.9   GROUND 121.7   CLEARANCE DELIVERY 127.25
APPROACH PHILADELPHIA 123.8   UNICOM 122.95   CTAF 126.9
ATIS 121.15.

Runway Lights   Tower and pilot-controlled lighting.

Transportation

Rental Cars
Avis, (800) 831-2847
Enterprise, (215) 281-3767
Hertz, (800) 654-3131
National, (800) 227-7368

Taxicabs
Manhattan Limousine, (800) 798-3671
United, (215) 238-9500
Yellow, (215) 922-8400

Courtesy Cars   North Philadelphia Jet Center has one available.

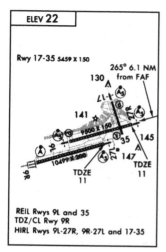

ELEV 22

Rwy 17-35 5459 x 150

265° 6.1 NM
130 from FAF

141

9500 X 150
35 145

10499 X 200
147
TDZE
11

TDZE
11

REIL Rwys 9L and 35
TDZ/CL Rwy 9R
HIRL Rwys 9L-27R, 9R-27L and 17-35

Airport Philadelphia International (PHL) is located 5 miles southwest of the city. Coordinates: N39° 52.22' W075° 14.70'.

Traffic Pattern The traffic pattern is flown at 821 ft MSL. Lefthand pattern for Runways 9L, 9R, 17, 27L, 27R, and 35. Runways 9L and 27R are asphalt surfaced, 9500 ft long by 150 ft wide. Runways 9R and 27L are asphalt surfaced, 10,499 ft long by 200 ft wide. Runways 17 and 35 are asphalt surfaced, 5459 ft long by 150 ft wide.

FBO Atlantic Aviation. (215) 492-2970. Hours: 24. Frequency 122.95.

Navigational Information PHL is located on the Washington Sectional Chart or L24, L28 low en route chart. From the DQO 114.0 VOR 065° at 20.3 miles.

Instrument Approaches ILS, VOR/DME, NDB, ASR, and GPS approaches are available.

Cautions There is a 320-ft lighted crane 3 nautical miles east of the airport. Birds and deer are in the vicinity.

Fuel 100LL and JetA are available. Texaco and most major credit cards are accepted.

Frequencies
TOWER 118.5   GROUND 121.9   CLEARANCE DELIVERY 118.85
DEPARTURE   119.75   APPROACH   PHILADELPHIA   128.4,
PHILADELPHIA 126.6   UNICOM 122.95   ATIS 133.4.

Runway Lights Operated by the tower.

Transportation

Rental Cars

Alamo, (215) 492-3960

Avis, (215) 492-0990

Budget, (215) 492-9447

Dollar, (215) 492-2692

Taxicabs

Carey Limousine Service, (215) 492-8402

Courtesy Cars   None available.

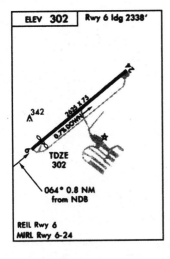

Airport   Wings (N67) is located 3 miles northwest of the city. Coordinates: N40° 08.19' W075° 16.02'.

Traffic Pattern   The traffic pattern is flown at 1500 ft MSL. Lefthand pattern for Runways 6 and 24. Runways 6 and 24 are asphalt surfaced, 2625 ft long by 75 ft wide.

FBO   Pennsylvania Aviation, Inc. (215) 646-1800. Hours: 24. Frequency 123.00.

Navigational Information   N67 is located on the New York Sectional Chart or L24, L28 low en route chart. From the MXE 113.2 VOR 064° at 22.8 miles.

Instrument Approaches   NDB and VOR/DME approaches are available.

Cautions   Large flocks of geese and deer are in the vicinity.

Fuel   100LL and JetA are available. Exxon and most major credit cards are accepted.

Frequencies
CLEARANCE DELIVERY 118.55   APPROACH PHILADELPHIA 126.85
UNICOM/CTAF 123.0.

Runway Lights   Operated by the tower.

Transportation

Rental Cars
Manhattan Limousine, (800) 798-3671
U Save, (215) 628-2100

Taxicabs
Radio, (215) 275-9700

Courtesy Cars   The FBO has one available.

Airport   Allegheny County (AGC) is located 4 miles southeast of the city. Coordinates: N40° 21.26' W079° 55.81'.

Traffic Pattern   The traffic pattern is flown at 2252 ft MSL. Lefthand pattern for Runways 5, 10, 13, 23, 28, and 31. Runways 10 and 28 are concrete surfaced, 6500 ft long by 150 ft wide. Runways 13 and 31 are asphalt surfaced, 3824 ft long by 100 ft wide. Runways 5 and 23 are asphalt surfaced, 2534 ft long by 100 ft wide.

FBO   Corporate Air Management, Inc. (412) 469-6800. Hours: 24. Frequency 122.95.

FBO   Corporate Jets, Inc. (412) 466-2500. Hours: 24. Frequency 122.95.

Navigational Information   AGC is located on the Detroit Sectional Chart or L23, L24 low en route chart. From the AGC 110.0 VOR 053° at 6.8 miles.

Instrument Approaches   ILS, VOR, NDB, VOR/DME, and GPS approaches are available.

Cautions   Birds and deer are in the vicinity.

Fuel   100LL and JetA are available. Air BP, Exxon, and most major credit cards are accepted.

Frequencies
TOWER 121.1   GROUND 121.7   APPROACH PITTSBURGH 119.35
UNICOM 122.95   ATIS 120.55.

Runway Lights   Operated by the tower.

Transportation

Rental Cars
Avis, (412) 462-2800

Taxicabs
For service call (412) 466-6066, 466-0166, 833-3300, or 466-9003.

Courtesy Cars   The FBOs have one available.

Area Attractions

Academy of Natural Sciences Museum.   19th St. and Benjamin Franklin Pkwy. (215) 299-1000. See dinosaurs, Egyptian mummies, animal displays, and more. Admission charged.

Betsy Ross House.   239 Arch St. (215) 627-5343. Visit the house where the famous seamstress is believed to have made the first American flag. Upholsterer's shop and memorabilia. Free.

Norman Rockwell Museum.   6th and Sansom Sts. (215) 922-4345. Drawings, prints, sketches, and more covering 60 years of Rockwell's career. Admission charged.

U.S. Mint. Independence Mall at 5th and Arch Sts. (215) 597-7350 or (202) 283-2646. Visitors are allowed an elevated view of the coinage operations and medal making. Coins of all denominations are produced here. Free.

Other Attractions

Colonial Mansions. Located within Fairmount Park. (215) 684-7922. Tour numerous 18th-century mansions in various architectural styles authentically preserved and furnished. Admission charged at all homes.

Independence National Historical Park. (215) 597-8974. Visit "America's most historic square mile" park. A tour map and information on all park activities and attractions can be picked up at The Visitor Center at 3d and Chestnut Sts. All historic sites and museums in the park are free.

Philadelphia Zoological Garden. 3400 W. Girard Ave. (215) 243-1100. See more than 1700 animals at America's first zoo. Admission charged.

Interesting Facts & Events

Mid-February. Jazz Weekend. Enjoy jazz concerts around the city. (215) 636-1666.

Begins Memorial Day weekend. Devon Horse Show. Horse Show Grounds in Devon, off U.S. 30. Nine days of horse competitions. (610) 964-0550.

Late November. Thanksgiving Day Parade. Come out for giant floats and celebrities. (215) 636-1666.

Early December. Fairmount Park Historical Christmas Tours. Enjoy period decorations in 18th-century mansions. (215) 684-7922.

For More Information Contact the Visitor's Center of the Philadelphia Convention and Visitor's Bureau at 16th St. and John F. Kennedy Blvd., (215) 636-1666 or (800) 537-7676, for more information, a tourist package, and discounts.

Lodging

Unless otherwise noted, all lodging is within 30 minutes of the primary airport.

Best Western Independence
Park Inn
235 Chestnut St.
(215) 922-4443
Rates: $99–$155

Days Inn–Airport
4101 Island Ave.
(215) 492-0400
Rates: $108–$118

Doubletree Club Hotel
9461 Roosevelt Blvd.
(215) 671-9600
Rates: $119–$175

Doubletree Guest Suites
4101 Island Rd.
(215) 365-6600
Rates: $119–$165

Holiday Inn Express Midtown
1305 Walnut St.
(215) 735-9300
Rates: $105–$135

Radisson
500 Stevens Dr.
(610) 521-5900
Rates: $135–$165

Restaurants

Chanterelles
1312 Spruce St.
(215) 735-7551
Specializes in rack of lamb.
Dinner: $22–$55

Ciboulette
200 S. Broad St.
(215) 790-1210
Specializes in Maine lobster.

Cutters–Commerce Square
2005 Market St.
(215) 851-6262
Specializes in fresh seafood.
Dinner: $12–$25

The Garden
1617 Spruce St.
(215) 546-4455
Specializes in aged prime beef.
Dinner: $17–$25

Rhode Island

Block Island

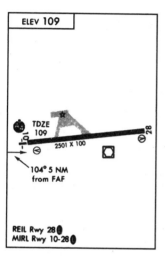

ELEV 109

TDZE
109

2501 X 100

28

104° 5 NM
from FAF

REIL Rwy 28
MIRL Rwy 10-28

Block Island is Rhode Island's summer resort that covers 21 mi² and lies 12 miles out to sea from Point Judith. It received its nickname as an Air-Conditioned Resort because it is 10 to 15 degrees cooler than the mainland in the summer. It is also milder in the winter.

Airport Block Island State (BID) is located 1 mile west of the city. Coordinates: N40° 10.09' W71° 34.67'.

Traffic Pattern The traffic pattern is flown at 1105 ft MSL. Lefthand pattern for Runways 10 and 28. Runways 10 and 28 are asphalt surfaced, 2501 ft long by 100 ft wide.

FBO New England Airlines. (401) 466-5881. Hours: 7:30 A.M.–8:30 P.M. Saturday through Thursday, 7:30 A.M.–10:30 P.M. Friday. Frequency 122.95.

Navigational Information BID is located on the New York Sectional Chart or L25, L28 low en route chart. The SEY 117.8 VOR is on the field.

Instrument Approaches VOR, VOR/DME, and NDB approaches are available.

Cautions There is a hill west and towers northeast. Deer and birds are in the vicinity.

Fuel 100LL and JetA are available. Most major credit cards are accepted.

Frequencies
CLEARANCE DELIVERY 120.1   APPROACH PROVIDENCE 125.75,
BOSTON CENTER 124.85   UNICOM/CTAF 123.0   AWOS 134.775.

Runway Lights Tower and pilot-controlled lighting.

Transportation

Rental Cars
Econo-Car, (401) 466-2029
Thrifty, (401) 466-2631

Taxicabs
Seacrest, (401) 466-2882
Your, (401) 466-5550

Courtesy Cars None available.

Area Attractions

Mohegan Bluffs. These 185-ft clay cliffs offer a sea view.

North Light. (401) 466-2982. Located near Settler's Rock, this light-house built in 1867 now functions as a maritime museum. Free.

Other Attractions

Fishing. (401) 466-2982. Surf casting is allowed from all beaches. There are freshwater ponds for bass, pickerel, and perch, and deep-sea boat trips for tuna, swordfish, etc.

For More Information Contact the Visitor's Center at (401) 466-2982 or (800) 383-2474 for more information, a tourist package, and discounts.

Lodging

Unless otherwise noted, all lodging is within 30 minutes of the primary airport.

Spring House
52 Spring St.
(401) 466-5844 or
(800) 234-9263
Rates: $149–$250

The Surf
Dodge St.
(401) 466-2241
Rates: $45–$130

Bed & Breakfast

The 1661 Inn and Guest House
1 Spring St.
(401) 466-2063
Rates: $115–$250

Restaurants

Finn's
Water St.
(401) 466-2473
Specializes in seafood and lobster.
Dinner: $5–$35

Mohegan Café
Water St.
(401) 466-5911
Specializes in seafood.
Dinner: $9–$19

Newport

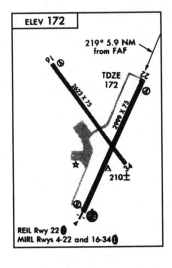

ELEV 172

219° 5.9 NM
from FAF

TDZE
172

2623 X 75

2899 X 75

210±

REIL Rwy 22
MIRL Rwys 4-22 and 16-34

Newport was founded by men and women fleeing religious intolerance in Massachusetts. Previously a large shipbuilding center, Newport is now known for its boating and yachting.

Airport Newport State (UUU) is located 2 miles northeast of the city. Coordinates: N41° 31.95' W071° 16.89'.

Traffic Pattern The traffic pattern is flown at 1170 ft MSL. Lefthand pattern for Runways 16 and 34. Runways 16 and 34 are asphalt surfaced, 2623 ft long by 75 ft wide.

FBO Northstar Aviation. (401) 738-2600. Hours: 24. Frequency 122.95.

Navigational Information UUU is located on the New York Sectional Chart or L25, L28 low en route chart. From the PVD 115.6 VOR 164° at 13.3 miles.

Instrument Approaches LOC, VOR/DME, and GPS approaches are available.

Cautions Birds and deer are in the vicinity.

Fuel 100LL and JetA are available. Most major credit cards are accepted.

Frequencies
CLEARANCE DELIVERY 127.25   APPROACH PROVIDENCE 128.7, BOSTON CENTER 124.85   UNICOM/CTAF 122.8.

Runway Lights Tower and pilot-controlled lighting.

Transportation

Rental Cars
Avis, (401) 846-1843
Hertz, (401) 846-1645
Rent-a-Car, (401) 846-3250
Thrifty, (401) 846-4371

Taxicabs
Aristocab, (401) 849-5454
Cozy, (401) 846-2500
Yellow, (401) 846-1500

Courtesy Cars   None available.

Area Attractions

Historic Mansions and Houses. Contact the Visitor's Bureau at (401) 849-8048 or (800) 326-6030, ext. 123, for ticket information to tour numerous historical mansions and homes. Admission charged at each home.

International Tennis Hall of Fame and Museum.   194 Bellevue Ave. (401) 849-3990. This is the world's largest tennis museum. Admission charged.

Old Colony and Newport Railroad.   America's Cup Ave. (401) 624-6951 or (401) 683-4549. Take a 1-hour scenic train ride along Narragansett Bay. Admission charged.

Other Attractions

Cliff Walk.   Begins at Memorial Blvd. Take a scenic walk overlooking the Atlantic Ocean.

The Newport Aquarium.   Easton's Beach. (401) 849-8430. Learn about native sea life. Admission charged.

Interesting Facts & Events

Late January–early February. Newport Winter Festival. 23 America's Cup Ave. Enjoy 10 days of food, music, cultural and recreational events, and activities. (401) 849-8048 or (800) 326-6030.

Mid-July. Newport Music Festival. Enjoy daily concerts in Newport's mansions. (401) 847-7090.

For More Information Contact the Newport County Convention and Visitor's Bureau at 23 America's Cup Ave., (401) 849-8048 or (800) 976-5122, ext. 123, for more information, a tourist package, and discounts.

Lodging

Unless otherwise noted, all lodging is within 30 minutes of the primary airport.

Courtyard by Marriott
9 Commerce Dr.
(401) 849-8000
Rates: $99–$225

Howard Johnson Lodge
351 W. Main Rd.
(401) 849-2000
Rates: $59–$268

Inn on the Harbor
359 Thames St.
(401) 849-6789
Rates: $110–$215

West Main Lodge
1359 W. Main Rd.
(401) 849-2718
Rates: $30–$75

Bed & Breakfast

Francis Malbone House
392 Thames St.
(401) 846-0392
Rates: $165–$325

Hammett House
505 Thames St.
(401) 848-0593
Rates: $99–$165

Mill Street Inn
75 Mill St.
(401) 849-9500
Rates: $125–$253

The Willows
8 Willow St.
(401) 846-5486
Rates: $88–$185

Restaurants

Canfield House
5 Memorial Blvd.
(401) 847-0416
Specializes in fresh seafood.
Dinner: $14–$23

Le Bistro
19 Bowen's Wharf
(401) 849-7778
Specializes in Angus beef.
Dinner: $10–$27

The Mooring
Sayer's Wharf
(401) 846-2260
Specializes in seafood.
Dinner: $9–$21

Providence

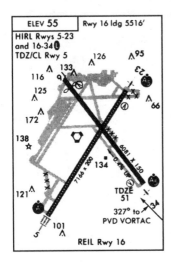

ELEV 55   Rwy 16 ldg 5516'
HIRL Rwys 5-23
and 16-34
TDZ/CL Rwy 5

327° to
PVD VORTAC

REIL Rwy 16

Providence is the capital and largest city of the state. Beginning as a farming community, it became a great industrial center in the 19th century.

Airport Theodore Francis Green State (PVD) is located 6 miles south of the city. Coordinates: N41° 43.41' W071° 25.60'.

Traffic Pattern The traffic pattern is flown at 1256 ft MSL. Lefthand pattern for Runways 16 and 34. Runways 16 and 34 are asphalt surfaced, 6081 ft long by 150 ft wide. Runways 5 and 23 are asphalt surfaced, 7166 ft long by 200 ft wide.

FBO Corporate Air Charter. (401) 732-0782. Hours: 8 A.M.–6 P.M. daily.

Navigational Information PVD is located on the New York Sectional Chart or L25, L28 low en route chart. The PVD 115.6 VOR is on the field.

Instrument Approaches ILS, ILS/DME, VOR, VOR/DME, NDB, and GPS approaches are available.

Cautions Sea gulls and deer are in the vicinity.

Fuel 100LL and JetA are available. Most major credit cards are accepted.

Frequencies
TOWER 120.7   GROUND 121.9   CLEARANCE DELIVERY 126.65
APPROACH PROVIDENCE 135.4, BOSTON CENTER 124.85
UNICOM 122.95   CTAF 120.7   ATIS 124.2.

Runway Lights Tower and pilot-controlled lighting.

Transportation

Rental Cars Available at the terminal.
Avis, (401) 738-5810
Budget, (401) 739-8660
Dollar, (401) 739-8450
Hertz, (401) 738-7500
National, (401) 737-4800
Thrifty, (401) 739-8660

Taxicabs
Airport, (401) 737-5550
Airport Van Shuttle, (401) 736-1900
Cozy, (401) 846-2500

Courtesy Cars The FBO has one available.

Area Attractions

The Arcade. Westminster St. Enjoy indoor shopping at over 35 specialty shops with national landmark statues.

Brown University. College St. (401) 863-1000. Brown is the seventh oldest college in the United States.

Culinary Archives and Museum. 315 Harborside Blvd. (401) 598-2805. Over 200,000 items of culinary art. Admission charged.

First Baptist Church in America. 75 N. Main St. (401) 454-3418. Oldest Baptist congregation in America. Free.

Museum of Rhode Island History at Aldrich House. 110 Benevolent St. (401) 331-8575. Headquarters for Rhode Island Historical Society. Admission charged.

Rockefeller Library. College and Prospect. (401) 863-2167. Library houses collections. Free.

Other Attractions

Lincoln Woods State Park. South of Breadneck Hill Rd. on RI 146. (401) 723-7892. Enjoy swimming, fishing, boating, hiking, bridle trails, ice-skating, picnicking, and more on over 600 acres. Admission charged for some activities.

Museum of Natural History. Elmwood Ave. (401) 785-9450. Educational arts programs. Admission charged.

Zoo. Elmwood Ave. (401) 785-3510. Children's nature center and educational programs. Admission charged.

Interesting Facts & Events

Spring. Festival of Historic Houses. (401) 831-7440. Tour selected private houses and gardens.

For More Information Contact the Providence Warwick Convention and Visitor's Bureau at (401) 274-1636 or (800) 233-1636 for more information, a tourist package, and discounts.

Lodging

Unless otherwise noted, all lodging is within 30 minutes of the primary airport.

The Biltmore
Kennedy Plaza
(401) 421-0700
Rates: $129–$750

Days Hotel
220 India St.
(401) 272-5577
Rates: $80–$125

Marriott
1 Orms St.
(401) 272-2400
Rates: $99–$250

Bed & Breakfast

Old Court Inn
144 Benefit St.
(401) 751-2002
Rates: $110–$250

States House Inn
43 Jewett St.
(401) 351-6111
Rates: $69–$109

Restaurants

Adesso
161 Cushing St.
(401) 521-0770
Specializes in pasta dishes.
Dinner: $10–$22

Al Forno
577 S. Main St.
(401) 273-9760
Specializes in grilled pizza
appetizers.
Dinner: $36–$45

Pot Au Feu
44 Custom House St.
(401) 273-8953
Specializes in classic and regional
French dishes.
Dinner: $12–$29'

Barre/Montpelier

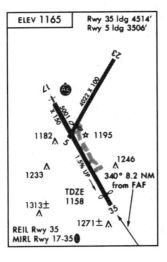

ELEV 1165 | Rwy 35 ldg 4514'
Rwy 5 ldg 3506'

1182
1195
1246
1233
340° 8.2 NM from FAF
TDZE 1158
1313±
1271±
REIL Rwy 35
MIRL Rwy 17-35

Barre was settled in 1788. Many highly skilled stonecutters have settled here because Barre has the world's largest granite quarries and granite-finishing plant. Ski resorts dot the area for winter sports enjoyment. Montpelier was settled in 1787. It is the state capital, a life insurance center, and a popular winter and summer vacation area.

Airport Edward F. Knapp State (MPV) is located 3 miles west of the city. Coordinates: N44° 12.21' W072° 33.74'.

Traffic Pattern The traffic pattern is flown at 2200 ft MSL. Lefthand pattern for Runways 5, 17, 23, and 35. Runways 17 and 35 are asphalt surfaced, 5001 ft long by 150 ft wide. Runways 5 and 23 are asphalt surfaced, 4022 ft long by 100 ft wide. Runways 5 and 23 are not plowed in the winter.

FBO Vermont Flying Services, Inc. (803) 223-2221. Hours: 9 A.M.– 5 P.M. daily. Frequency 122.8.

Navigational Information MPV is located on the Montreal Sectional Chart or L20 low en route chart. From MPV-110.8 VOR 340° at 8.1 miles.

Instrument Approaches ILS, VOR, VOR/DME, and NDB approaches are available.

Fuel 100LL is available at the FBO. Most major credit cards are accepted.

Frequencies
APPROACH BOSTON CENTER 135.7   UNICOM/CTAF 122.8   ASOS
132.675.

Runway Lights Pilot-controlled lighting.

Transportation

Rental Cars
Avis, (802) 229-5922
Barre Ford, (802) 479-0136
Budget, (802) 476-4724
Capital Chrysler, (802) 479-0586
Hertz, (802) 223-3815 or (800) 645-3131

Taxicabs
A&D, (802) 476-9408
Barre, (802) 479-1857
Garrett's, (802) 479-2127
Norms, (802) 223-5226

Courtesy Cars None available.

Area Attractions

Barre

Granite Sculptures. Hope Cemetery. On VT 14 at the north edge of
town. Granite sculpture of headstones carved by artisans as a final tribute
to themselves and their families.

Robert Burns. Downtown near the city park. Regarded as one of the
world's finest granite sculptures.

Youth Triumphant. Benches around the memorial create a whisper
gallery.

Groton State Forest. VT 232, near Groton. (802) 584-3822. Elevation
1078 ft. Includes 3-mile-long Lake Groton and six other ponds. Hiking
trails for the enthusiast, nature trails with picnicking, and swimming facilities available.

Montpelier

Hubbard Park. Hubbard Park Dr. (802) 223-5141. Stone observation tower located on 110-acre wooded park with picnic area. Free.

Morse Farm. 3 miles north via County Rd. (802) 223-2740. Maple sugar and vegetable farm in rustic wooded setting. Free.

State House. Constructed of Vermont granite in 1859 with gold leaf covered dome. Free.

Other Attractions

Barre

Rock of Ages Quarry and Manufacturing Division. Main St. in Graniteville. (802) 476-3119. Tour the quarry and manufacturing division as skilled artisans create monuments. Free.

Montpelier

Thomas Waterman Wood Art Gallery. Vermont College Arts Center at College St. (802) 828-8743. Monthly changing exhibits of oils, watercolors, and etchings by Thomas Waterman Wood.

Vermont Historical Society Museum and Library. Pavilion Office Building next to State House. (802) 828-2291. Historical exhibits.

Interesting Facts & Events

Last weekend in September. Old Time Fiddlers' Contest. Barre Auditorium. (802) 476-0256.

For More Information Contact the Central Vermont Chamber of Commerce at (802) 229-4619 or (802) 229-5711 for more information, a tourist package, and discounts.

Lodging

Unless otherwise noted, all lodging is within 30 minutes of the primary airport.

Barre

Days Inn
173 S. Main St.
(802) 476-6678
Rates: $39–$85

Hollow Inn
278 S. Main St.
(802) 479-9313
Rates: $80–$99

Montpelier

Comfort Inn at Maplewood Ltd.
VT 62
(802) 229-2222
Rates: $69–$79

La Gue Inns
VT 62 near E. F. Knapp Airport
(802) 229-5766
Rates: $50–$80

Bed & Breakfast

Barre

Green Trails
I-89 Exit 5
(802) 276-3412
Rates: $75–$161

Montpelier

Betsey's Bed & Breakfast
74 E. State St.
(802) 229-0466
Rates: $50–$75

Inn on the Common
Main St., Craftsbury Common
(802) 586-9619
Rates: $135–$240

The Inn at Montpelier
147 Main St.
(802) 223-2727
Rates: $89–$149

Restaurants

Barre

Country House
276 N. Main St.
(802) 476-4282
Specializes in seafood, veal, and pasta.

Montpelier

Chef's Table
118 Main St.
(802) 229-9202
Continental menu.

Lobster Pot
313 Barre
(802) 476-9900
Specializes in seafood and steak.

Burlington

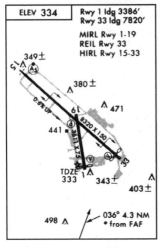

ELEV 334	Rwy 1 ldg 3386' Rwy 33 ldg 7820'
	MIRL Rwy 1-19 REIL Rwy 33 HIRL Rwy 15-33

Located on Lake Champlain, Burlington is the largest city in Vermont. Settled in 1773, it is the birthplace of Ethan Allen.

Airport Burlington International (BTV) is located 3 miles east of the city. Coordinates: N44° 28.29' W073° 09.17'.

Traffic Pattern The traffic pattern is flown at 1335 ft MSL. Lefthand pattern for Runways 1, 15, 19, and 33. Runways 15 and 33 are asphalt surfaced, 7807 ft long by 150 ft wide. Runways 1 and 19 are asphalt surfaced, 3611 ft long by 75 ft wide.

FBO Innotech Aviation. (802) 658-2200. Hours: 6 A.M.–10 P.M. daily. Frequency: 122.95.

FBO Valet Air Services. (802) 863-3626 or (800) 782-0773. Hours: 6 A.M.–9 P.M., 24 hr on request. Frequency 123.5.

Navigational Information BTV is located on the Montreal Sectional Chart or L26 low en route chart. From BTV 117.5 VOR 036° at 4.3 miles.

Instrument Approaches ILS, VOR, NDB, ASR, and GPS approaches are available.

Cautions Visibility is obscured at the approach end of Runway 15 by an industrial plant. Birds are in the vicinity.

Fuel 100LL and JetA are available. Exxon and most major credit cards are accepted.

Frequencies
TOWER 118.3   GROUND 121.9   APPROACH BURLINGTON 121.1
UNICOM 122.95   ATIS 123.8   ASOS 123.8.

Runway Lights   Operated by the tower.

Transportation

Rental Cars
Avis, (802) 864-0411
Budget, (802) 658-1211
Hertz, (802) 864-7409
National, (802) 864-7441

Taxicabs
Commencers, (802) 863-1889

Courtesy Cars   Innotech Aviation has one available.

Area Attractions

Battery Park.   VT 127 and Pearl St. View Lake Champlain and the Adirondacks. Free.

Bolton Valley Ski/Summer Resort.   I-89 Exits 10, 11. (802) 434-2131. This resort offers all the amenities of summer and winter sports. Admission charged.

Ethan Allen Homestead.   Off VT 127. (802) 865-4556. Preserved pioneer homestead of the 1787 farmhouse. Guided tours are available. Admission charged.

Excursion Cruises.   Burlington Boathouse, College St. (802) 862-8300. Cruise Lake Champlain in a replica of a vintage sternwheeler.

Lake Champlain Chocolates.   431 Pine St. (802) 864-1807. View the art of making chocolate.

Other Attractions

Green Mountain Audubon Nature Center.   I-89 Richmond Exit. (802) 434-3068. There are 230 acres with trails through many environments to view beaver ponds, hemlock swamps, brooks, old farm fields, woodlands, and a sugar orchard.

University of Vermont. Fifth oldest university in New England offering graduate and undergraduate programs. Built in 1791.

Interesting Facts & Events

Three days in June. Lake Champlain Balloon and Craft Festival. Balloon launches, rides, crafts, and entertainment.

For More Information Contact the Lake Champlain Regional Chamber of Commerce at (802) 863-3489 for more information, a tourist package, and discounts.

Lodging

Unless otherwise noted, all lodging is within 30 minutes of the primary airport.

Bel-Aire
111 Shelburne St.
(802) 863-3116
Rates: $55–$85

Comfort Inn
1285 Williston Rd.
(802) 865-3400
Rates: $65–$99

Holiday Inn
1068 Williston Rd.
(802) 863-6363
Rates: $83–$129

Howard Johnson
1720 Shelburne Rd.
(802) 860-6000
Rates: $80–$120

Sheraton Hotel and
Conference Center
870 Williston Rd.
(802) 865-6600
Rates: $99–$145

Restaurants

Amigo's
1900 Shelburne Rd.
(802) 985-8226
Specializes in Mexican food.

Daily Planet
15 Center St.
(802) 862-9647
Specializes in rack of lamb.

Ice House
171 Battery St.
(802) 864-1800
Specializes in fresh seafood.

Sirloin Saloon
1912 Shelburne Rd.
(802) 985-2200
Specializes in wood grilled steak.

Pauline's
1835 Shelburne Rd.
(802) 862-1081
Specializes in fresh seafood.

Sweetwaters
120 Church St.
(802) 864-9800
Specializes in bison burgers.

Nova Scotia

Halifax

Halifax is the capital of Nova Scotia and the largest city in the Atlantic provinces.

Airport   Halifax International (CYHZ) Coordinates: N44° 52.51' W063° 30.31'.

Traffic Pattern   The traffic pattern is flown at 1477 ft MSL. Lefthand pattern for Runways 6, 15, 24, and 33. Runways 6 and 24 are asphalt surfaced, 8800 ft long by 200 ft wide. Runways 15 and 33 are asphalt surfaced, 7700 ft long by 200 ft wide.

Instrument Approaches   ILS, NDB, and VOR/DME approaches are available.

Fuel   100LL and JetA are available.

Frequencies
TOWER 118.4   CLEARANCE DELIVERY 123.95   GROUND 121.9
ATIS 121.0.

Area Attractions

Art Gallery of Nova Scotia.   1741 Hollis-at-Cheapside. (902) 424-7542. Changing exhibits. Free on Tuesdays.

Maritime Museum of the Atlantic.   Lower Water St. (902) 424-7490. Exhibits of nautical history. Admission charged.

Nova Scotia Museum of Natural History.   1747 Summer St. (902) 424-7353. Exhibits on Nova Scotia. Admission charged.

Other Attractions

Murphy's on the Water.   Murphy's Pier next to Historic Properties. (902) 420-1015. Sightseeing tours. Admission charged.

Province House.   Hollis St. (902) 424-8967. Oldest Parliament building in Canada.

Interesting Facts & Events

February.   Winterfest.

July.   Highland Games.

September.   Nova Scotia Air Show.

For More Information   Contact the International Visitor's Centre at 1595 Barrington St., (902) 490-5946 or (800) 565-0000, for more information, a tourist package, and discounts.

Lodging

Unless otherwise noted, all lodging is within 30 minutes of the primary airport.

Airport Hotel
60 Bell Blvd.
(800) 667-3333
Free airport transportation.
Rates: $97–$144

Inn on the Lake
Hwy. 102
(800) 463-6465
Free airport transportation.
Rates: $97–$275

Days Inn
636 Bedford Hwy.
(902) 443-3171
Rates: $65–$89

Stardust
1067 Bedford Hwy.
(902) 835-3316
Rates: $35–$55

Restaurants

Five Fishermen
1740 Argyle St.
(902) 422-4421
Specializes in seafood.
Dinner: $9–$15

Salty's on the Waterfront
1869 Upper Water St.
(902) 423-6818
Specializes in lobster and steak.
Dinner: $14–$22

Ontario

Ottawa

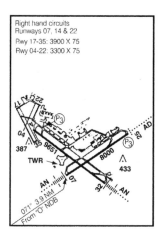

Right hand circuits
Runways 07, 14 & 22
Rwy 17-35: 3900 X 75
Rwy 04-22: 3300 X 75

Ottawa is the capital city of Canada.

Airport Macdonald–Cartier International (CYOW) Coordinates: N45° 19.21' W075° 40.09'.

Traffic Pattern The traffic pattern is flown at 1374 ft MSL. Lefthand pattern for Runways 4, 25, and 32. Righthand pattern for Runways 7, 14, and 22. Runways 4 and 22 are asphalt surfaced, 3300 ft long by 75 ft wide. Runways 7 and 25 are asphalt surfaced, 8000 ft long by 200 ft wide. Runways 14 and 32 are asphalt surfaced, 9651 ft long by 200 ft wide.

Instrument Approaches ILS, NDB, and VORTAC approaches are available.

Fuel 100LL and JetA are available.

Frequencies
TOWER 118.8, 120.1   CLEARANCE DELIVERY 119.4   DEPARTURE 128.175   ATIS 121.15, 132.95   ARR 135.15.

Area Attractions

Canadian Museum of Civilization. 100 Laurier St. (819) 776-7000. Permanent and changing exhibits. Admission charged.

Canadian War Museum. 330 Sussex Dr. (819) 776-8627. Exhibits of Canada's military history. Admission charged.

Currency Museum. 245 Sparks St. (613) 782-8914. Exhibits of money and its use. Admission charged.

National Aviation Museum. Rockcliffe Airport. (613) 993-2010. More than 100 historic aircraft. Admission charged.

National Gallery of Canada. 380 Sussex Dr. at St. Patrick St. (613) 990-1985. Arts, prints, drawings, photos, and film. Free.

National Museum of Science and Technology. 1867 St. Laurent Blvd. (613) 991-3044. More than 400 exhibits. Admission charged.

Other Attractions

Ottawa Riverboat Company. (613) 562-4888. Cruises on the Ottawa River.

Ottawa Senators. 1000 Palladium Dr. (613) 721-0115. Professional hockey.

Royal Canadian Mint. 320 Sussex Dr. (613) 991-5853. Production of coins. Admission charged.

Interesting Facts & Events

Three weekends in February. Winterlude. Outdoor concerts and skating contests.

July 1. Canada Day. Celebrate Canada's birthday. (613) 239-5000.

For More Information Contact the tourism and Convention Authority at 130 Albert St., (613) 237-5150, for more information, a tourist package, and discounts.

Lodging

Unless otherwise noted, all lodging is within 30 minutes of the primary airport.

Best Western Hotel
Jacques Cartier
131 Laurier St.
(819) 770-8550
Rates: $86–$125

Radisson–Ottawa Centre
100 Kent St.
(613) 238-1122
Rates: $125–$400

Sheraton
150 Albert St.
(613) 238-1500
Rates: $175–$390

Bed & Breakfast

Voyageur's Guest House
95 Arlington Ave.
(613) 238-6445
Rates: $34–$44

Restaurants

Al's Steak House
3817 Richmond Rd.
(613) 828-8349
Specializes in steak.
Dinner: $11–$36

Marble Works
14 Waller St.
(613) 241-6764
Specializes in seafood.
Dinner: $12–$18

The Mill
555 Ottawa River Pkwy.
(613) 237-1311
Specializes in prime rib.
Dinner: $10–$16

Toronto

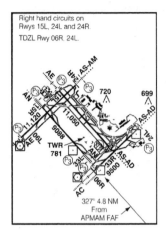

Right hand circuits on Rwys 15L, 24L and 24R.

TDZL Rwy 06R, 24L.

720

699

TWR

781

327° 4.8 NM
From
APMAM FAF

Toronto is a leading industrial, commercial, and cultural center for Canada.

Airport Lester B. Pearson International (CYYZ) Coordinates: N43° 40.38' W079° 37.50'.

Traffic Pattern The traffic pattern is flown at 1569 ft MSL. Lefthand pattern for Runways 6L, 6R, and 33. Righthand pattern for Runways 15, 24L, and 24R. Runways 6L and 24R are asphalt surfaced, 10,500 ft long by 200 ft wide. Runways 6R and 24L are concrete surfaced, 9500 ft long by 200 ft wide. Runways 33 and 15 are asphalt surfaced, 11,050 ft long by 200 ft wide.

Instrument Approaches ILS, NDB, and VOR/DME approaches are available.

Fuel 100LL and JetA are available.

Frequencies
TOWER 118.35, 118.7   GROUND 121.9, 121.65   CLEARANCE DELIVERY 121.3   DEPARTURE 128.8, 127.575   ATIS 112.15, 114.8, 120.825   ARR 125.4, 124.475.

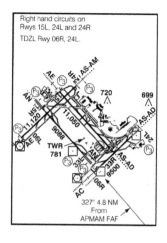

Right hand circuits on Rwys 15L, 24L and 24R.
TDZL Rwy 06R, 24L.

327° 4.8 NM
From
APMAM FAF

Airport Toronto/City Centre (CYTZ) Coordinates: N43° 37.39' W079° 23.46'.

Traffic Pattern The traffic pattern is flown at 1251 ft MSL. Righthand pattern for Runways 6, 8, and 15. Lefthand pattern for Runways 24, 26, and 33. Runways 6 and 24 are asphalt surfaced, 3000 ft long by 150 ft wide. Runways 8 and 26 are asphalt surfaced, 4000 ft long by 150 ft wide. Runways 15 and 33 are asphalt surfaced, 3000 ft long by 150 ft wide.

Instrument Approaches A NDB approach is available.

Fuel 100LL and JetA are available.

Frequencies
TOWER 118.2, 119.2   GROUND 121.7   DEPARTURE 133.4   ATIS 133.6   ARR 133.4   CTAF 118.2.

Area Attractions

Art Gallery of Ontario.   317 Dundas St. W. (416) 977-0414. Exhibits of paintings, drawings, and sculpture. Free on Wednesday evenings.

Casa Loma.   1 Austin Terr. (416) 923-1171. A medieval-style castle. Admission charged.

CN Tower.   301 Front St. W. (416) 360-8500. Three observation decks. Admission charged.

Metro Toronto Zoo.   Meadowvale Rd. (416) 392-5900. Explore 710 acres featuring native and exotic animals. Admission charged.

Ontario Science Centre.   770 Don Mills Rd. (416) 429-4100. Exhibits of space, technology, and earth science. Admission charged.

Other Attractions

Toronto Blue Jays. 1 Blue Jays Way. (416) 341-1000. Major League baseball.

Toronto Maple Leafs. 60 Carlton St. (416) 977-1641. Professional hockey.

Toronto Raptors. 1 Blue Jays Way. (416) 214-2255. Professional basketball.

Toronto Stock Exchange. King and York Sts. (416) 947-4676. Free.

Woodbine Racetrack. Hwy. 427. (416) 675-RACE. Thoroughbred racing. Admission charged.

Interesting Facts & Events

Mid-July. Outdoor Art Show. (416) 408-2754.

Mid-August–Labor Day. Canadian National Exhibition. Events include parades, water shows, and air shows. (416) 393-6000.

For More Information Contact Tourism Toronto at 207 Queens Quay W., (416) 203-2500 or (800) 363-1990, for more information, a tourist package, and discounts.

Lodging

Unless otherwise noted, all lodging is within 30 minutes of the primary airport.

Crowne Plaza–Toronto Centre
225 Front St. W.
Rates: $219–$435

Days Inn–Downtown
30 Carlton St.
(416) 977-6655
Rates: $89–$129

Hilton
145 Richmond St. W.
(416) 869-3456
Rates: $239–$1600

Marriott–Airport
901 Dixon Rd.
(416) 674-9400
Free airport transportation.
Rates: $240–$1000

Quality–Airport East
2180 Islington Ave.
(416) 240-9090
Near airport.
Rates: $85–$135

Ramada Hotel—Toronto Airport
2 Holiday Dr.
(416) 621-2121
Free airport transportation.
Rates: $140–$350

Restaurants

Centro Grill
2472 Yonge St.
(416) 483-2211
Specializes in rack of lamb.
Dinner: $19–$31

Scaramouche
1 Benvenuto Pl.
(416) 961-8011
French menu.
Dinner: $25–$30

Chiaro's
King St. Hotel
37 King St. E.
(416) 863-9700
Specializes in steak.
Dinner: $23–$38

360 Revolving Restaurant
301 Front St. W. in CN Tower
(416) 362-5411
Specializes in prime rib.
Dinner: $25–$36

About The Author

Douglas S. Carmody, CFII, has been flying the country for 25 years. A commercial airline pilot since 1987, he is a graduate of Embry-Riddle Aeronautical University and holds a Gold Seal CFI, CFII, and MEI instructor ratings. Owner of a flight training and education business, he captains a Boeing 737 for USAirways and is the author of the *McGraw-Hill Pilot Test Guide Series.*

Your Cockpit Guide to

Perfect for tucking into a flight bag, this one-of-a-kind travel guide not only tells you what to expect when flying in—it shows you where to find the most fun, the best food, the choicest lodgings, and the best pilot discounts after landing!

This guide gives you pilot-tested information as well as useful tips on:

- AIRPORTS—traffic patterns, runways, hours, navigational info, instrument approaches, fuels, costs, cautions, radio frequencies, runway length, and more
- ATTRACTIONS—historic sites, museums, and unique offerings, with admission fees; recreational and shopping opportunities; interesting local notes, color, and history
- LODGINGS—accommodations from low cost to luxury, with phone numbers
- FOOD AND DINING—including specialties, prices, and distances from the airport, from quick, decent burger joints to restaurants worthy of being destinations themselves
- RENTAL CARS—including firms that will pick you up and deliver you back to the airport
- DISCOUNTS JUST FOR PILOTS—more than enough to pay for this book!

Don't miss these other Pilot's Trave & Recreatia Guides:

Northwest and Western Canad

Southeast and Caribbean

Southwest and Baja

In addition to Eastern Canada, this guide covers: Connecticut, Delaware, Illinois, Indiana, Maine, Massachusetts, Michigan, New Hampshire, New Jersey, New York, Ohio, Pennsylvania, Rhode Island, and Vermont

Cover Design: Susan Newman • Cover Photo: Tony Stone Images

ISBN 0-07-001743-3

Visit us on the World Wide Web at
www.books.mcgraw-hill.com/aviation

McGraw-Hill
A Division of The McGraw-Hill Companies

$24.95 U.S.A.

9780070017436